THE PERFECT THEORY

A Complete Unified Description of the Universe

BY

GEOFFREYSON KHAMALA

i

Published By:

FOUNDATION

"Thinking for the Universe"

DEDICATION

I dedicate this book to thinking gurus, their admirers and critics worldwide.

TABLE OF CONTENTS

LIST OF ACRONYMS & ABBREVIATIONS

BCE   –   Before Common Era

CE   -   Common Era

DNA   –   Deoxyribonucleic Acid

EPR paradox -   Einstein–Podolsky–Rosen paradox

etc.   -   et cetera☐

HST   -   Hubble Space Telescope

i.e.   -   that is

NGOs   -   Non-Governmental Organizations

QCD   -   Quantum Chromo-Dynamics

ToE   -   Theory of Everything

PUBLICATIONS BY GEOFFREYSON KHAMALA

1. The Perfect Theory: A Complete Unified Description of the Universe (2014)

2. What is science? Science as an Adaptive Capacity (2014)

3. Is Science Religion? (2014)

4. Wither Globalization Enter Connectedness (2014)

5. Tajiriba Spaces: The solution to Sub-Optimal Outcomes (2015)

6. Zero Unemployment in Kenya: The Utility of Tajiriba Spaces (2015)

7. Reclaiming the Sahara: A Case for Universal Connectedness (2015)

ABSTRACT

Ever since the dawn of humanity, humans have attempted to fully comprehend and explain nature. However, orthodox science holds that not even the most sophisticated model is proficient enough to provide a perfect depiction of the universe. Indeed, no thought process thus far has ever managed to present social life, politics, economics, psychology, technology, the material world and terrestrial space as a seamless whole. Atomists described nature in the form of atoms and the surrounding void (empty space). Newtonian mechanics comprehended the natural world as being made up of particles and absolute space. Albert Einstein while explaining gravity suggested that the universe is best grasped as spatial (space) and temporal dimensions (time) – space-time. Today all the conventional fundamental interactions of nature, except gravity, are explained using the standard theory of particle physics. Albert Einstein's theory of gravitation and the Standard Model are known to be inconsistent with each other. Whereas the quantum theory describes the micro world (the subatomic composition of the universe such as electrons and quarks), the general theory of relativity excellently describes the macro world (cosmological scales such as stars, planets and galaxies). Einstein realized that the way out of the quagmire is to discover a Theory of Everything (ToE) that would unify or explain through a single model all the fundamental interactions of nature. Einstein died before achieving this feat. Following Einstein, a number of scientists have attempted to originate theories of everything (denoted as quantum gravity) such as string theory, superstring theory, M-theory and loop quantum gravity. However, critics observe that these theories do not constrain the characteristics of the predicted universe(s) and/or are inconsistent with observations.

Without a doubt Albert Einstein nearly cracked the code of the universe when he attempted time travel. 'Travelling' as fast as or faster than the speed of light defines the interface between Einstein's General Relativity and Quantum Mechanics. On the other hand, religious traditions discovered this reality countless millennia before Einstein. Science started with religion. The idea of supernatural forces, gods and God marks the first step in human awareness that they are not condemned to be mortals forever. So, religious entrepreneurs were the first humans to imagine perpetuity. However, religion is limited in the quest by humans to become immortal because of the separation of heavenly and earthly realms. Contemporary mainstream scientists explicitly reject divine explanations for human occurrence, search for natural cause, and base their conclusions on evidence. Unlike religious rituals, the conventional scientific tradition is such that other scientists confirm and improve on existing knowledge. However, science as a method cannot enable us to discover and describe wholly the underlying nature of reality. Therefore, to explain the universe as a single unifying whole it is important to ascertain why we study and/or practice science. What defines science, evolution and experience is the purpose not the method or any other criterion.

The Perfect Theory is a dialogue on the nature of reality with the world's foremost scientists. The theory almost perfectly explicates the nature of the universe in a single theoretical framework by demonstrating that the subject matter of science is spatial (non-Euclidian space) and non-spatial dimensions (inherent properties of

vii

the physical world). We are one with nature. The physical world is extended in non-physical realms in what can be described as connectedness. Hence, in addition to the familiar three–dimensional spatial world, there are twelve (12) non-spatial relations namely gravity, electromagnetism, strong nuclear interaction, weak nuclear interaction, dark matter, dark energy, mental faculties, senses, emotions, time, life and death. Spatial and non-spatial dimensions are the reference frames in which we organize our experience. Following Immanuel Kant, the physical (*noumenal*) world cannot be comprehended directly. The subject for scientific study is the phenomenal world (phenomena or properties of the physical world). We grasp the universe by studying non-physical phenomena.

The gist of *The Perfect Theory* is that connectedness is what drives science and everything else. Connectedness as a thought process represents another noteworthy stride in the human quest for meaning. Connectedness is essential to our understanding of the origins, structure and ultimate fate of the universe, and our place in it. The universe is an organic whole, a whole in which all parts are connected. We are one (share a purpose) regardless of the supposed diversity.

History, evolution and our life experience are entangled. History is the development of human awareness of our connectedness with the rest of the universe. The centerpiece of the connectedness experience is life (and death).

Life (and death) is the organizing principle of the universe. The human society, for example, is constituted by bonds of hope (life) and bonds of fear (death). Humankind is self-centered, competitive and conflictual out of fear (the urge for self-preservation). However, broadly understood identity struggles and politics in the human society are not attempts to gain access to and control limited resources - the social (e.g., mates), biological (e.g., food), and physical (e.g., territory) rather humans are engaged in the pursuit to perpetuate life and eschew bereavement (collective preservation).

Since the dawn of humankind, human beings have desired to live forever. Mourning is the most conspicuous collective expression of the human longing to circumvent death. As scientists (and the general public) perfect their mastery of the completeness of the cosmos (spatial and non-spatial dimensions), humanity may possibly manage to travel back in time, explore distinctive futures and effectively become immortal.

Keywords: The Ultimate Theory, Physical Phenomena, Non-physical Phenomena, mourning, immortality

CHAPTER ONE

FUNDAMENTALS OF THE UNIVERSE

Fundamental Principles

Studies dating back to the prehistoric period have been motivated by the quest to develop an uncomplicated model of the universe. The desire to fully explain the universe has been one of the driving forces behind the elusive but crucial efforts in science to develop a comprehensive explanation of nature (or the Theory of Everything). *The Perfect Theory* is an attempt to reveal the underlying order to the seeming arbitrariness of the physical world. Virtually everything can be accommodated in its unrestrained embrace. Underpinning *The Perfect Theory* is the understanding that:

i. All things (living and non-living) are interdependent and inseparable parts of the cosmic whole;
ii. Opposites cohere;
iii. Change occurs at the margins;
iv. The principle(s) that governs natural phenomena also governs human behavior and social patterns; and
v. The principle of the universe is to sustain life.

All is One; One is All

From early on inquiring minds have strongly believed that the nature of the universe is solitary but complex. For instance, according to Plato (423-347 BC) in *Parmenides* "You cannot conceive the many without the one.... The study of the unit is

1

among those that lead the mind on and turn it to the vision of reality". For Alexander Pope "all are but parts of one stupendous whole". Fritjof Capra in the *The Tao of Physics: An Exploration of the Parallels Between Modern Physics and Eastern Mysticism* (first published in 1975) observes that the teachings of ancient Eastern philosophies such as Hinduism, Buddhism, Taoism, Shintoism and Zen are in agreement with recent discoveries in science that although many phenomena of the observable world are seemingly unrelated, they all go back to the same principle. Archimedes of Syracuse (287 BCE – 212 BCE) the greatest mathematician of all time summed it all: "Give me a place to stand, and I will move the Earth" (Mackay, 1991). To paraphrase Archimedes, the world is characterized by oneness rather than separate parts.

The properties of space have been postulated by many, from ancient Vedic philosophy, Eastern Mystics, various ancient civilizations throughout human history all the way to Newton, Descartes, Einstein, and more.

Michael S. Schneider in *A Beginner's Guide to Constructing the universe: The Mathematical Archetypes of Nature, Art and Science* (1995) observes that the "discovery and appreciation of the circle is our early glimpse into the wholeness, unity, and divine order of the universe". To be sure, the circle is a metaphor for unity and wholeness. This is because the circle accommodates all shapes within itself. Apparently the circle has neither an end nor a beginning. The same thought process is apparent in numerous scientific attempts to discover the laws that govern nature namely mind-body duality, Newtonian gravity, Kant's categorical imperative, Quantum

2

Mechanics, Cartesian geometry, Einstein's space-time and infinity. Religion also gave the world the idea of eternity.

Things are intertwined and interdependent. Thus, in the multiplicity of things there is unity. The physical world is many things and one thing at the same time. Accordingly, accurate awareness of the unity and mutual interrelation of all things and events makes us aware they represent different manifestations of the same ultimate reality. Connectedness is a new way to render scientific knowledge. Connectedness summarizes our understanding of nature. Connectedness merges two previously incompatible fields: natural sciences and human sciences.

Unity of Opposites

The way we describe the universe is dominated by unities of opposites as denoted by difference and sameness. The general principle of unity of opposites submits to a situation in which every actual thing or situation involves a coexistence of opposite conditions, yet subject on each other and presupposing each other. Accordingly, a perfect understanding of how and why these pairs of opposites cohere holds the key to understanding the universe.

Table of Examples of Pairs of Opposites

Connectedness	Disconnectedness	The Perfect Theory
Reality	Appearance	Socrates, Plato & Parmenides
Rest/gravitation	Time/change/motion/acceleration	Heraclitus, Parmenides & Einstein
Something	Nothing	Parmenides
Atoms/matter	Void/Vacuum/Emptiness	Democritus & Leucippus
Particles/matter	Absolute space	Newtonian Mechanics
Spatial dimensions	Temporal (time) dimensions	Einstein's Space-Time Continuum
Spatial dimensions	Non-spatial dimensions	The Perfect Theory
Body	Mind	Rene Descartes
Forms	Ideas	Plato
Sensible world	Intelligible world	Plato
Apparent world (changes)	Unseen and unchanging world of forms	Plato
Wisdom/knowledge	Ignorance	Socrates, Confucius, Tao
Secular wisdom	Divine wisdom	Jesus
Contact forces	Action at a distance	Isaac Newton
Up/ The road up	Down/ The road down	Heraclitus
Abundance	Scarcity	The Perfect Theory
Unemployment	Full employment/inflation	Philips Curve
Supply	Demand	Alfred Marshall
Deflation	Inflation	The Perfect Theory
Fixed exchange rate	Floating exchange rate	A Single World Currency
Monetary	Fiscal	Perfect Economy
Immigrants	Emigrants	Labour Mobility
Brain gain	Brain drain	Brain Circulation
Natural or pure sciences	Human sciences	Science
North	South	Comparative
West	East	Comparative
Rich	Poor	Comparative
Life (immortality)	Death (mortality)	Perfect Life
Natural Phenomena	Artificial (human) events	Connectedness
Nature	Nurture	Connectedness
Earth/hell	Sky/heaven	Jesus of Nazareth
City of man	City of God	St Augustine
Sasa	*Zamani*	John Mbiti
Evolution/science/experience	Big bang	The Perfect Theory
Abundance	Scarcity	The Perfect Theory
Life	Non-life/death	Ancients; Peter Higgs/the God Particle/Higgs boson
Life	Death	The Perfect Theory
Action	Reaction	Isaac Newton
Expansion	Contraction	Expansion Theory
Matter	Antimatter	Paul Dirac
One (1)	Zero (0)	Multiple states – Quantum computers
God	Gods	Science/Immortality
Tension/tensional stress	Compression/compression stress	

4

Science as an Academic Endavour and As a Practice

Religious and philosophical speculation about the physical world long predates contemporary scientific inquiry in attempts to grasp the astoundingly vast and old universe. Mention can be made of the creation story that is told in the book of Genesis which can be traced to a Babylonian epic of the 22nd century BCE. The motivation for science is to understand the underlying reality behind the appearances. Science as a field of study has grown in leaps and bounds. However, this journey of growth has been characterized by specialization such that today if the scientific community wanted to take of the stock of knowledge thus far it becomes an extremely herculean task. It is unmistakably obvious that a lot has been done in science as evidenced by material, technical and technological progress in the world. Now and again, scientists have come forward with completely inventive ideas that have contribute to and modified how we view the universe. Nevertheless, throughout the years scientists have not succeeded to bring together known knowledge.

Today validated knowledge is scattered under different specializations and sub-disciplines. However, too much specialization distorts reality such that we are unable to answer the most basic questions such as: Why do we do science? You may say we study science in the quest to know but is mere knowing enough? More importantly what connects science as a whole? Such a noble endeavor require a universal principle and subsequently a super theory (a grand synthesis) — a theory that will reconcile all the sciences (natural, human, social, formal and other sciences) with the everyday activities (or events) in the physical world.

The guiding principle of *The Perfect Theory* is that the subject matter of science (both as an academic endeavor and as a practice) is one. Connectedness is the sum of all human knowledge.

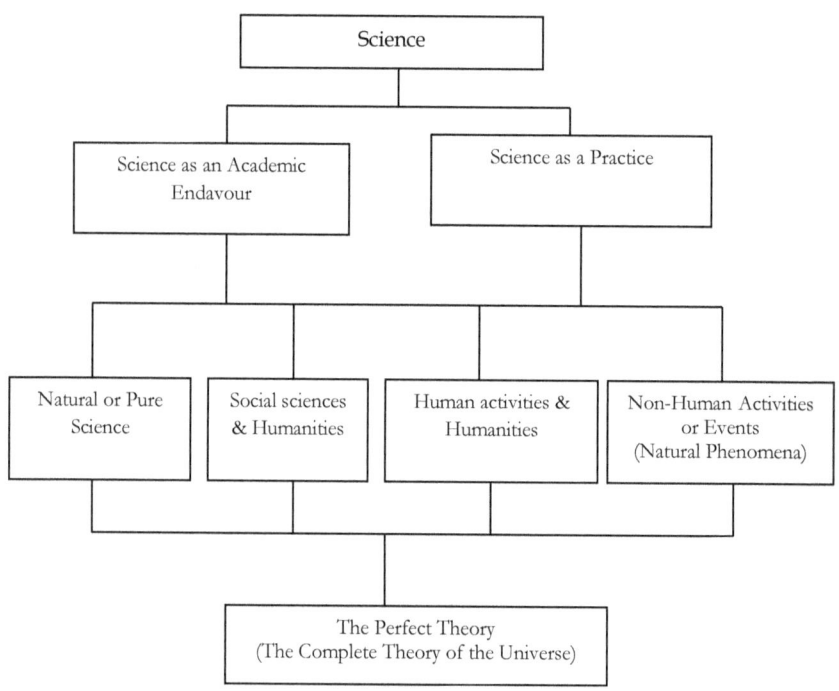

Figure 1: Schematic Representation of Science in the Holistic Sense

CHAPTER TWO

THE GRAND SYNTHESIS: SCIENCE AND ACTIVITIES IN THE PHYSICAL WORLD

Atoms viz-as-viz Empty Space

The sticking point is whether everything changes (Heraclitus) or nothing ever changes (Parmenides). The challenge is whether everything is infinitely divisible (Anaxagoras) or the supposition that everything in the universe is either atoms or voids (Leucippus and Democritus atomic theory). Today scientists argue that there is only one fundamental particle in the universe, the electron. Of course this sharply contrasts with the idea of subatomic particles in nature like protons, neutrons and quarks. Still, according to quantum theory electrons (and light) exhibit characteristics of both waves and particles.

At the heart of these debates is whether we can grasp the universe. Is the universe intelligible? Parmenides, Socrates and Plato understood that appearances can be deceiving. Socrates associated knowledge with the apprehension of unchanging eternal forms. Plato distinguished between forms and ideas posturing two distinct worlds: the sensible world and the intelligible world. He believed in a higher reality of which the material world is just a manifestation. For Plato the material world as it seems to us is not the real world, but only an image or copy of the real world which can only be perceived by reason. Plato seems to distinguish two worlds: the apparent world, which constantly changes, and an unchanging and unseen world of forms, which may be the cause of what is apparent.

For Parmenides (early 5th century BCE) all change is an illusion; the true underlying reality is eternally unchanging and of a single nature. In particular, he noted the ability of matter to change forms (e.g., ice to water to steam). This made Parmenides and a number of his contemporaries to propose that all the apparently different materials of the world are different forms of a single primordial material ranging from water, air, mind, fire, earth, and wind. Following Parmenides, Democritus and Leucippus proposed nature exists only of two things, namely atoms and the void that surrounds them. At the most fundamental level atomism is the explanatory scheme based on the understanding that all phenomena are explicable in terms of the properties and behaviour of ultimate, elementary, localized entities (or indestructible atoms) separated by the emptiness of the void.

Following atomism, it was suggested that all forces could be ultimately reduced to contact forces between the indivisible atoms then imagined as tiny solid particles moving through the void (vacuum). A vacuum is a space devoid of atoms, the units that make up air, other gases and familiar objects. That means a vacuum is the next best thing to a space truly empty of anything at all. According to these early atomists, nature exists only of two things, namely atoms and the emptiness that surrounds them. Atomists speculated that the apparent diversity of observed phenomena was due to rearrangements of the atoms as a consequence the collisions of atoms. By this time, the idea that forces could be mediated by a vacuum was well accepted. But there were three major setbacks for thinking this way.

First, Parmenides said empty space is fiction, because it is impracticable to describe something that is not anything. This view was accordingly summarized: "Nature abhors a vacuum". Likewise, Aristotle did not believe that the void was feasible because air, like water, offers resistance to motion. Air will immediately rush in to fill a void. We now know that space can be without solid matter, but it always contains some form of radiation. In fact, new evidence indicates that it is possible to extract light from seeming emptiness (vacuum). Further, hydrogen can be created from space under certain energetic conditions.

Secondly, Aristotle (384 BC – 322 BC) originated the notion that the universe is bounded by a collection of 55 concentric celestial spheres (or celestial orbs), the outermost being heaven, surrounding those of the planets, earth and its elements, and finally the seven circles of hell. He suggested that the concentric spheres rotate around the earth with the assistance of a god – an unchanging divine mover.

Finally, Isaac Newton's description of the long-distance force of gravity implied that the idea of exclusively contact forces in nature had to be discarded. Prior to Newton, people didn't think carefully about the implications of gravity. No one had ever seen an object exert a force on another object except when the objects were in contact. Newton's theory of universal gravitation proposed exactly that.

Particles viz-as-viz Absolute Space

Newtonian Mechanics

In ancient times, Aristotle hypothesized that objects of different masses fall at different rates. But during the Scientific Revolution, Galileo Galilei (1564 – 1642CE) experimentally determined that this was not the case. He realized that neglecting the friction due to air resistance, all objects accelerate toward the Earth at the same rate. That is, mass equals to weight. The understanding of nature underwent radical transformations when Isaac Newton discovered the mechanical model of the universe. According to the Newtonian mechanics all physical phenomena take place in an absolute three dimension space of classical Euclidean geometry. Newton pioneered the idea that every particle in the universe attracts another particle with some force, whether in space or on Earth: "what goes up, must come down". The apparent attracting force in nature that attracts material objects together instead of repelling them away from one another he labeled gravity. According to his work *Philosophiæ Naturalis Principia Mathematica* (1687), the force of gravity acting on an object is also that object's weight (not mass). Mass is not quite the same as weight. Mass is defined as a given quantity of matter regardless of the weight. Weight is related to the pull of gravity on an object. Gravity behaves differently in outer space. The mass of an object would be the same on earth as it would on the moon which has much less gravity.

Newton demonstrated that the same invisible force (gravity) that caused apples to fall to the ground was responsible for the movements of the moon. Algebra and geometry were useful for describing the size of stationary objects. But Newton sought to describe things that are moving or changing in some way. He invented calculus so that he could grasp motion in time (moving targets) particularly of the sun, the moon, the planets and the stars in the night sky.

A force is a push or pull upon an object resulting from the object's interaction with another object. The four fundamental forces are gravity, the electromagnetic force, and the weak and strong nuclear forces. All forces (interactions) between objects can be placed into two broad categories: contact forces, and forces resulting from action-at-a-distance. Contact forces are those types of forces that result when the two interacting objects are perceived to be physically contacting each other. Examples of contact forces include frictional forces, tensional forces, normal forces, air resistance forces, and applied forces. Action-at-a-distance forces are those types of forces that result even when the two interacting objects are not in physical contact with each other, yet are able to exert a push or pull despite their physical separation. Examples of action-at-a-distance forces include gravitational forces, magnetic force and electrical force. The sun and planets, for example, exert a gravitational pull on each other despite their large spatial separation. Still, the protons in the nucleus of an atom and the electrons outside the nucleus experience an electrical pull towards each other despite their small spatial separation. Finally, two magnets can exert a magnetic pull on each other even when separated by a distance of a few centimeters.

The theoretical description of material bodies in Newtonian physics was based on the material point or particle; matter was thus considered a priori to be discontinuous. Newton's theory of gravity was founded on the motion of discrete particles in absolute space and time, in the sense of being quantifiable objects such as chair and dog. Leibniz suggested that time and space are not material things. Newton was, therefore, faced with the dilemma of explaining how matter (tiny particles) could interact with other matter at a distance across the universe (in space). He assumed that gravitation acts instantaneously, regardless of distance. Gravity was treated as a case of action-at-a-distance - the interaction of two objects which are separated in space with no known mediator of the interaction. Accordingly, each particle with mass responds instantaneously to every other particle with mass irrespective of the distance between them. Therefore, the theory assumes the speed of gravity to be infinite. But Newton guided by rationalism (clear causes and effects) denied the possibility for a body to act upon another at a distance through a vacuum without the mediation of anything else. For him, action-at-a-distance is unfeasible.

The collapse of the Newtonian world view (absolute space and time, elementary solid particles and problem of action-at-a-distance) contributed to the emergence of Quantum Mechanics and Einstein's general theory of relativity. For Einstein, gravitation is not force acting at a distance as Newton had suggested but the outcome of the curvature of space-time. This means that gravity can be explained in terms of geometry, rather than as interacting forces. The carrier (or messenger) particle of gravity is the graviton. The graviton has not been experimentally demonstrated, mainly because it is awfully difficult to find the smallest denomination of the weakest interaction. Latest postulations indicate that it will likely be massless.

Standard Model/Quantum Mechanics

The contemporary understanding of the subatomic composition of the universe (the nature of matter) is summarized in what is known as the standard theory of particle physics. The Standard Model (or Quantum Mechanics/quantum theory) describes both the fundamental building blocks out of which the world is made, and the forces through which these blocks interact, that is, the behavior of matter and its interactions with energy on the scale of atoms and atomic particles. Material objects are made up of discrete tiny particles of energy (quanta) and the forces that keep those objects together. Particles are considered to be points moving through space and are distinguished by their position and velocity. The universe is made up of six particles including Higgs boson ("God particle"), which is critical to understanding the structure of the universe. The Higgs boson is thought to give mass to other particles and is so short-lived that it can only be detected by the evidence it leaves.

In 1900, Max Planck, began the Standard Model of particle physics to explain atomic and subatomic processes. He postulated that everything is made up of little bits of matter called quanta (one quantum; two quanta). Later, in 1905, Albert Einstein through his theory of the photoelectric effect (which won him the Nobel prize for physics) posited that light could exist in discrete particle-like quantities (photons) contrasting the classical view of light as a continuous wave. Planck's and Einstein's theories were progenitors of Quantum Mechanics. Quantum Mechanics deals with the knowledge of the universe beyond the general perception of matter by our ordinary senses of tasting, seeing, hearing, feeling, and

13

sensing. Matter has both wave and particle aspects. Major contributors to Quantum Mechanics besides Max Planck and Albert Einstein include Niels Bohr, Wolfgang Pauli, Erwin Schrödinger, Werner Heisenberg, Max Born, Paul Dirac and Louis de Broglie.

In 1913, Niels Bohr discovered the semi-classical model of the atom. The Bohr model depicts the atom as a small, positively charged nucleus surrounded by electrons that travel in circular orbits around the nucleus but with attraction provided by electrostatic forces rather than gravity. Atoms (or particles) are deemed to be points moving through space and are uniquely identified by their position and velocity. At one time, scientists contemplated whether two particles in the same quantum state could exist in the same place at the same time. In 1923, Wolfgang Pauli discovered the Pauli Exclusion Principle observing that a certain group of particles (bosons) do not obey this principle. Particles that do obey the Pauli Exclusion Principle are called fermions. The Pauli Exclusion Principle holds that no two identical fermions can exist in identical energy quantum state simultaneously. No two electrons in an atom can have identical quantum numbers (i.e. no two electrons in an atom can be at the same time in the same state or configuration). The Pauli Exclusion Principle is part of one of our most basic observations of nature and helps explain a wide variety of physical phenomena such as the observed patterns of light emission from atoms.

In 1925, Schrödinger discovered the wave function. The wave function is a statistical depiction used to estimate the prospect that the particle is in a particular location or in a state of motion. The wave function is the heart of Quantum

Mechanics. The wave function is a probability function used by physicists to understand the nanoscale world. At the sub-atomic scale, there is no boundary between particles and waves. The Schrodinger equation is used to find the allowed energy levels of quantum mechanical systems (such as atoms, or transistors). The associated wave function gives the probability of finding the particle at a certain position. Schrödinger's equation gave a correct description of an electron's behavior (electron energies) in almost all cases. However, physically interpreting the very meaning of the waves is one of the main philosophical problems of Quantum Mechanics.

When a measurement of the wave/particle is made, its wave function collapses. Wave function collapse is the phenomenon characterized by the reduction of the physical possibilities into a single possibility as seen by an observer. For instance, light acts both like a wave and a particle. The wave/particle duality at first seemed paradoxical until physicists managed to assimilate this subtle marriage of opposites into their mathematics. Using the wave function, physicists can calculate a system's future behavior. However, the reduction of the physical likelihoods into a single likelihood as seen by an observer has always been debated. For some, this prospect represents an observer's subjective state of knowledge of the universe. Subsequently, wave function collapse corresponds to the receipt of new information about some underlying reality but does not represent reality. For others, ranging from Schrödinger, Einstein, Bohm and Everett etc., the wave function really exists and the collapse of the wave function is also a real process that objectively exists whether or not an observer is observing or measuring it.

In 1927, Werner Heisenberg introduced the Heisenberg Uncertainty Principle. According to Heisenberg, not all properties of a quantum particle can be measured with complete accuracy. The more precisely the position of a particle is determined, the less precisely the momentum is known simultaneously, and vice versa. He demonstrated that our observations have an effect on the behavior of quanta (discrete quantities of energy) such that it is impossible to measure anything without disturbing it. For example, when experimenters use light to observe the object being observed they can influence the behavior of particles changing energy of motion. In a similar fashion, the Copenhagen interpretation of Quantum Mechanics first posed by Niels Bohr and his colleagues at the University of Copenhagen suggests that a quantum particle doesn't exist in one state or another but in all of its possible states at once. When observed, an object is essentially forced to take one state or another (probability), and that's the state that we observe experimentally. Since it may take a different observable state each time, this explains why a quantum particle behaves erratically.

In 1927, Niels Bohr developed complementarity (wave-particle duality). In the quantum world everything has both a particle and a wave aspect. All properties of quantum entities exist only in pairs (conjugate pairs). Light, for instance, is besides that bright stuff which makes things visible, also a force (the so-called electromagnetic force) that keeps electrons tied to the nuclei of atoms, and atoms tied together to make molecules and finally objects. Bohr emphasized on the role played by observers in the process of measurement demonstrating that the object being measured is unavoidably affected by the measurement. He concluded that it is impossible to design a measuring device that demonstrates the behaviour of atomic objects and the interaction with the measuring instruments simultaneously simply because such a device is literally inconceivable. Differently put, Bohr suggests that it is not possible to regard objects governed by Quantum Mechanics as having intrinsic properties independent of determination with a measuring device.

In 1927, Max Born provided critical statistical interpretation of the wave function especially superposition or wave overlapping. Werner Heisenberg, Erwin Schrödinger and Paul Dirac formulated the basic quantum mathematics.

In 1935, Erwin Schrödinger noted a peculiar property he called entanglement. Hitherto almost all of physics was based on Einstein's principle of locality. The notion of local action holds that a particle influences another only by direct contact or via some intermediary field and this influence can travel no faster than light. Distant objects cannot have direct influence on one another. Non-locality, on the other hand, would mean that one particle could influence another distant particle without anything passing between them, in an instantaneous manner (faster-than-light). Nonlocality occurs due to the phenomenon of entanglement, whereby particles that interact with each other effectively lose their individuality and in many ways behave as a single entity. Schrödinger demonstrated that when two quantum systems are brought together and then separated, they remain still connected by an instantaneous new kind of wholeness.

Quantum Mechanics' reject the view that physical processes must obey local causes (locality) by demonstrating that action-at-a-distance is feasible. This position gained wide acceptance after the discovery of electric and magnetic fields by Faraday, Maxwell, and others. In Quantum Mechanics, the chemical bonds resulting from electrical interactions between the smallest particles of matter fall under electromagnetism. The theory of electromagnetism considers electricity, magnetism, and electromagnetic waves (e.g., light, radio waves, and X-rays) to be different aspects of a single phenomenon (the same fundamental interaction).

According to the theory of electromagnetism action at a distance can be accounted for the concept of a field which mediates interactions between currents and charges across empty space. The recently observed Higgs boson is generated by a field (the Higgs field) and permeates the vacuum of space giving mass to all particles of matter. According to the theory, all electromagnetic waves travel at the speed of light. A magnet, for instance, attracts a piece of iron directly through the intermediate empty space but the magnet always calls into being something physically real in the space around it - a "magnetic field". In its turn, this magnetic field operates on the piece of iron, so that the latter strives to move towards the magnet. Accordingly, electrostatic interactions between charged particles produce around themselves an electric field, which can be felt by other charges as a force.

As the 19th century came to a close, scientists thought that through the Standard Model they had captured an almost complete picture of the nature of the universe. For them, the physical world comprises of two basic entities: particles and fields. Particles are tiny bits of matter while fields such as electricity, gravity and magnetism are spread out over vast regions and when disturbed vibrate and travel as waves. However, at the beginning of the 20th century, in spite of the overwhelming success of Quantum Theory, a number of limitations became apparent.

First, Einstein, Podolsky, and Rosen (EPR paradox) demonstrated that the wave function does not provide a complete physical description of reality thereby suggesting that the entire theory of Quantum Mechanics was radically incomplete. It emerged that the Standard Model provides a fairly comprehensive scientific understanding of the universe particularly the workings of the universe at subatomic levels (electromagnetic, strong and weak forces). However, the most familiar force in our everyday lives, gravity, is not part of the Standard Model. For example, electromagnetism is infinite-ranged like gravity, but vastly stronger, and therefore describes a number of macroscopic, and many atomic level phenomena of everyday experience such as friction, elasticity, pressure, rainbows, lightning, and all human-made devices using electric current, such as television, lasers, and computers. Even though electromagnetism is far stronger than gravitation, electrostatic attraction is not relevant for large celestial bodies, such as planets, stars, and galaxies. Therefore, particle theory works well when gravity is so weak that it can be neglected or when we pretend gravity doesn't exist. Yet there exists a rather strong belief that gravitation may end up being the most important feature in the ultimate explanation of the universe.

Second, unlike electric and magnetic fields, objects under the sole influence of gravitational fields receive an acceleration, which does not in the least depend either on the material or on the physical state of the body.

Third, there exists the unresolved question of whether reality is local or non-local. Quantum holds that reality is non-local. The phenomenon of entanglement envisages correlations between remote events. Bell's Theorem (1964) demonstrated that spooky actions at a distance are unavoidable. Since then many experiments have confirmed these long-distance correlations. However, the EPR paradox shows that the phenomenon of entanglement seemingly violates the principles of local causality. Indeed, Albert Einstein scorned hypothetical non-local influences as "spooky actions at a distance".

Fourth, the inherently probabilistic nature of quantum theory has generated a nearly century-long debate on whether the wave function has an objective physical existence (represents objective reality) or merely the subjective knowledge of an observer. In 1927, Niels Bohr and others advocated the alternative view in the Copenhagen interpretation, in which the wave function is merely a mathematical probability that immediately assumes only one value when an observer measures the system, resulting in the wave function collapsing.

Finally, nature does not show us any isolated building blocks but rather appears as a complicated web of relations between the various parts of the whole raising doubts over the conception of elementary particles as the primary units of the universe (matter).

Albert Einstein helped develop Quantum Mechanics. But fitting gravity comfortably into the standard framework has proved to be a difficult challenge. Einstein and his colleagues (EPR Paradox) concluded that Quantum Theory could not be a final theory of nature. Indeed in contrast to the Einstein's General Relativity which is about the largest things in the universe, Quantum Mechanics deals with the interaction of atomic and subatomic particles, the particles that atoms are made of. Even so, John Stewart Bell opined that a single, elegant and straightforward theorem cannot happen. For him, no physical theory of local hidden variables can ever reproduce all of the predictions of Quantum Mechanics.

Spatial and Temporal Dimensions[1]

Space-Time

One of Euclid's most enduring legacies was his development of Euclidean flat space geometry. Euclid (330 B.C.) in *The Elements* noted that a point is dimensionless, a line is one-dimensional, a plane is two-dimensional and finally space is three-dimensional. Euclid believed that his axioms were self-evident statements about physical reality.

Today we describe our physical world as being three-dimensional thanks to the Cartesian/analytic geometry. Following Euclid, Descartes (1596-1650) discovered an abstract way of conceptualizing physical space using the right angle (the Cartesian plane). He used algebra to describe three-

---

[11] Dimensions are basically the distinctive aspects of what we make out to be reality as we know it.

dimensional geometry, where every point in Euclidean space is represented by an ordered triple of coordinates (x, y, z). Before Descartes, monists held that the mind is not something separate from the body; the mind is all that exists and the external world is either mental itself, or an illusion created by the mind. Descartes distinguished between the realm of mind and the realm of matter. He was the first to clearly identify the mind with consciousness and self-awareness. According Descartes, consciousness (the mind) can exist independently of physical reality (e.g. the brain). Accordingly, to appropriately grasp the universe one has to develop a few simple, fundamental ideas of reality from which geometrical deduction of more complex ideas and theories emerges.

Leibniz anticipated Albert Einstein by arguing, against Newton, that space, time and motion are relative, not absolute. Einstein invented differential geometry when he extended his Special Theory of Relativity to encompass Newton's theory of gravitation with the end result being *General Theory of Relativity* (1915).

Einstein adopted three-dimensional geometry to indicate physical space. But his theory shows that the true geometry of space-time is not Euclidean geometry. His understanding was that space is not nothing. Accordingly, the conceptualization of space detached from any physical content (matter, objects) does not exist. The physical reality of space is represented by a continuous field; the universe consists of particles of matter and energy located in and operating through space. He favored the situation where distance and time are measured by relative positions of objects in our universe. He observed that though our common sense tells us that space and time are fundamentally separate things, the fabric of space and time are actually linked. For Einstein, gravity is a distortion (curve) in the shape of space-time. Space-time is curved by the presence of matter. Gravity is so strong that space is bent round onto itself making it rather like the surface of the earth (the earth is spherical rather than flat). The effects of gravity depend on two things: the mass of two bodies and the distance between them. An implication of Einstein's theory of General Relativity is that Euclidean space is a good approximation to the properties of physical space only where the gravitational field is not too strong.

Einstein identified two kinds of dimensions, spatial (bidirectional) and temporal (unidirectional). Dimensions are independent properties of a coordinate grid needed to locate a point in a certain defined space. Accordingly, any real body must have extension in four directions: it must have length, breadth, thickness, and duration. Typically, to uniquely determine where and when events occur anywhere on the universe (rather than just points in space) you need specify space (i.e. the latitude and longitude) and the duration (time).

Whereas space has no boundary the universe is not infinite in space. Gravity is generated when matter causes space-time to curve like a heavy object (e.g. bowling ball) would stretch a rubber sheet then smaller objects roll downwards, towards the larger object. Gravity for Einstein is just another phrase for the curvature of space-time; a warping of our planet's surface. If we could travel in a certain direction on the surface of the earth hoping to reach the end of the universe, we could never come up against an impassable barrier or fall over the edge, but eventually come back to where we started. To bolster the argument that the surface of the earth is curved Aristotle in 340 BC argued that one always sees the sails of a ship coming over the horizon first and only later its hull.

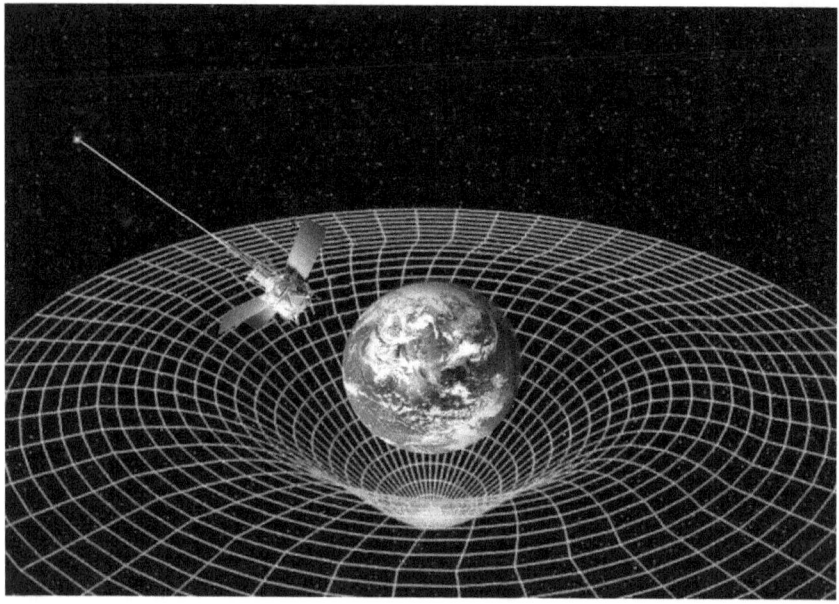

Figure 2: An Illustration of the space-time curvature caused by Earth

Source: Wikipedia

Einstein proposed that gravitation is responsible for various astronomical phenomena observed on Earth. If there were no gravity the universe would be flat and featureless. In everyday life, gravitation is most familiar as the agent that gives weight to objects with mass, causes bodies to fall to the ground when dropped and sees to it that animals can only jump so high. Gravity keeps us grounded on the planet, keep the moon in its orbit around the earth, prevents our solar system from flying apart by keeping the Earth and the other planets in their orbits around the Sun and binds together enormous clusters of galaxies. Gravitation is also responsible for causing dispersed matter to coalesce, and coalesced matter to remain intact, thus accounting for the existence of the earth, the sun, and most of the macroscopic objects in the universe.

Space and time appear to be changeable entities, which account for the constancy of the speed of light. Space and duration are one because the observed rate at which time passes for an object depends on the object's velocity relative to the observer and also on the strength of gravitational fields, which can slow the passage of time. The duration of time can therefore vary for various events and various reference frames. For instance, time slows down at higher speeds of the reference frame relative to another reference frame (time dilation). It is impossible for any particle that has rest mass to be accelerated to the speed of light. Nothing in the universe can travel faster than light. If you were capable of travelling faster than the speed of light you could reach your destination almost instantly. If you can travel faster than the speed of light then you would in fact be moving faster than what light can report to you, and hence, you could at least witness the past. When we look at the universe, we are seeing it as it was in the past. That is why we do not know what is happening at the moment farther away in the universe.

Einstein showed how both temporal and spatial dimensions can be altered (or "warped") by high-speed motion. Subsequently people travelling at different speeds, while agreeing on cause and effect, will measure different time separations between events and can even observe different chronological orderings between non-causally related events. Though these effects are typically minute in the human experience, the effect becomes much more pronounced for objects moving at speeds approaching the speed of light. The curvature of space-time is negligible on small scales. So, Einstein's general theory of relativity describes gravity as it pertains to the large-scale structure of the universe but falters when used to with phenomena on extremely small scales (particles separated by very short distances) such as contact forces, elasticity, viscosity, friction, and pressure.

General Relativity has yielded a wealth of insight into the universe, the orbits of planets, the evolution of stars and galaxies, the Big Bang, the expanding universe, black holes and gravitational lenses. However, the theory itself only works when we pretend that Quantum Mechanics is not needed in our description of nature.

Einstein resisted quantum theory because of its inherent uncertainty and its assertion that particles can remain linked even when separated by great distances (non-locality). In the general theory the highest possible speed for any physical interaction in nature is the speed of gravitational waves and is equal to the speed of light in vacuum. For Einstein, instantaneous action-at-a-distance violates the universe's ultimate velocity limit yet nothing can travel faster than the speed of light. Still, Quantum Mechanics appears to violate relativistic causality. Because of causality, cause precedes effect, or cause and effect may appear together in a single item, but effect never precedes cause.

General Relativity and the Standard Model are known to be inconsistent with each other. Whereas the quantum theory describes the micro world, the general theory of relativity describes the macro world. Einstein realized that the way out of the quagmire is to discover a theory of everything that would unify or explain through a single model all the fundamental interactions of nature: gravitation, strong interaction, weak interaction, and electromagnetism. Gravitation is extremely weak relative to the other four interactions. Yet it is long ranged and therefore relevant for understanding cosmological distances and the evolution of the universe. However, all the interactions except gravity are explained using quantum theory.

In most ordinary physical situations, gravity applies to super-galactic levels of the universe while Quantum Mechanics describes the smallest structures in the universe such as electrons and quarks. However, some situations require both, for instance, central point of a black hole or the state of the universe just before the big bang, the process scientists believe brought the known universe into being. This scenario necessitates the quest for the elusive but scientifically crucial Theory of Everything.

Expansion Theory

Edwin Hubble having studied the redshift of distant galaxies concluded that the universe is expanding. Hubble's observations imply that the cosmos had a beginning. Creation myths across traditions (Chinese, Indian, pre-Colombian, African, the biblical book of Genesis) also attach a beginning to the universe.

Saul Perlmutter, Brian Schmidt and Adam Riess were in 2011 awarded the Nobel Prize for discovering that the expansion of the universe was happening at an ever-increasing rate. Saul Perlmutter and his colleagues demonstrated that a careful study of the shifts of distant galaxies shows that the universe is actually accelerating its expansion.

Expansion Theory is an attempt to explain everything that previously required three theories (a blend of Newton's gravity, Quantum Mechanics and General Relativity). Until recently, scientists expected to see gravitational attraction between galaxies slowing down the expansion of the cosmos or ultimately force the universe into contraction. In a universe which is dominated by matter, one would expect gravity eventually should make the expansion slow down. But according to the three scientists (and supernovae studies), the universe is constantly expanding in all directions. Expansion theory predicts a non-static cosmos and avers that the universe is not only expanding but that its expansion rate is increasing. The theory predicts that the universe will expand forever at an even faster rate due to dark energy. Expansion theory says that the reason gravity is so indistinguishable from an acceleration is because it is acceleration, caused by matter expanding at an ever accelerating rate. The composition of the universe is 5% normal matter (protons, neutrons and electrons), 23% dark matter and 72% antigravity dark energy. Dark matter can only be detected by looking at the gravitational effects it has on the space around it (it cannot be sensed as such). Antimatter exists only as products of radioactive decay, exploding stars, and very strong electric fields.

Some scientists even suggested that it is possible that the universe is oscillating between eras of expansion and contraction. Conversely, according to expansion theory, the density of the attractive gravitational pull of matter and the repulsive gravitational push of dark energy are not constants. The density of matter decreases as the universe expands because the volume of space increases.

Dark energy and cosmic acceleration are a failure of General Relativity on very large scales. Standard theory assumes that the universe is static and eternal, and consists of a single galaxy, surrounded by a vast, infinite, dark, and empty space. Einstein's theory of General Relativity was a description of a perfectly static universe that had no beginning or end. However, gravity being a purely attractive force between all objects, it is impossible to have a set of masses located in space at rest forever. Their mutual gravitational attraction will ultimately cause them to collapse inward, in manifest disagreement with an apparently static universe. Some unknown force with negative pressure pervading space seems to be pushing the universe apart. The mysterious force that repels gravity is dark energy. Even then, it's likely that the accelerated expansion of the universe is an illusion caused by the relative motion of us to the rest of the universe. If this evidence holds up, then dark energy would not exist.

Candidate Theories of Everything

Scientists seek to elucidate every single aspect of the universe (gravity, electromagnetism, and strong and weak nuclear forces) in one theory. Theories of everything (denoted as quantum gravity) attempt to develop a grand theoretical/mathematical framework to describe all known observable fundamental interactions at both sub-atomic (Quantum Mechanics) and cosmological scales (stars, planets, galaxies). It is an attempt to quantize General Relativity. Quantum gravity theories include string theory, superstring theory, M-theory and loop quantum gravity.

The physicist Michio Kaku in the 1970s postulated string theory suggesting that small, vibrating strings serve as the building blocks of all matter, and that their vibrations create all four of the forces in our universe. All matter and energy is made of tiny vibrating strings. String theories have attempted to harmonize all the particles and forces of nature (including gravity) into a single theoretical framework. The objective is to successfully describe all fundamental forces and forms of matter. Typically the three spatial dimensions (length, width, height), and one temporal dimension (time) are required to describe in a more uniform way the workings of the universe. But according to string theories there are extra dimensions but we only see three space and one time dimension. String theories posit that the electrons and quarks within an atom are not 0-dimensional objects, but rather 1-dimensional oscillating lines ("strings").

The essential splendid idea behind string theories is this: all of the different fundamental particles are really just different manifestations of one basic object - a tiny vibrating string. The electron is not really a point, but a tiny loop of string. The entire world is made of strings. Matter and energy (what appear to us as particles) are composed of tiny vibrating strings (tiny spring-like loops) of various types that exist in a space-time with 10, or even 26 dimensions.

Superstring Theory suggests that the universe exists in ten (10) different dimensions. So, besides the three visible dimensions of extent (height), thickness (breadth) and deepness (depth) can be found seven extra dimensions which are not directly evident to us.

M-theory is an eleven-dimensional framework that is believed to encompass all of the previously five distinct superstring theories.

String theorists suggest that this modification of quantum field theory where point particles are replaced with string-like objects that propagate in space-time environment allows gravity and Quantum Mechanics to form a harmonious union.

Is locality (the fact/quality of having a position in space or the situation/position of an object) defined by 3, 4, 5, 6 11 or 26 dimensions? String theory predict 10 to 26 dimensions (with M-theory predicting 11 dimensions: 10 spatial and 1 temporal). Therefore, string theories seem to predict extra dimensions but the number of dimensions is not fixed by any consistent criterion. The implication is that how many dimensions are needed to describe the universe is still an open question. In fact, the existence of more than four dimensions would only appear to make a difference at the subatomic level. Further, finding a way to confirm string theory with our current technology is a major challenge. Further still, quantum gravity theories all assume, and to some degree depend upon, the existence of the graviton. Finally, given the aforementioned reasons no string theory has secured wide acceptance.

Robert Lanza and Robert Berman in *Biocentrism: How Life and Consciousness Are the Keys to Understanding the True Nature of the universe* (2009) developed a theory of biocentric universe (biocentrism). The theory builds on quantum physics with the hope of climaxing in a theory of everything. They argue that what we call space and time are not external physical objects but rather forms of sense perception. For them, life and consciousness are central to a comprehensive understanding of the universe.

Other proposed space-time theories include additional spatial and temporal dimensions or dimensions that are neither temporal nor spatial. For example, multiple universes theory imagines the existence of hypothetical multiverse (sometimes called parallel universes or parallel worlds).

The major predicament for these candidate theories of everything is that they do not constrain the characteristics of the predicted universe(s) or are inconsistent with observations. String theories create universes with arbitrary numbers of dimensions while multiple universes theory creates unseen dimensions that are outside normal experience, mysteriously beyond our normal three-dimensional comprehension. Still, the quest for a theory of everything has been confined discipline inbreeding. Scientists must interrogate events outside their discipline requiring collaboration and borrowing from other disciplines. Finally, a complete theory of the universe has to incorporate quantum theory (the Standard Model/Quantum Mechanics) and gravitation. Such a theory has not been found, yet.

The Perfect Theory: Spatial and Non-Spatial Dimensions

Underlying *The Perfect Theory* is the idea that the universe is **spatial** (space) and **non-spatial dimensions** (properties of the physical world). Herein, (non-Euclidian) space is the conventional way of looking at the universe based on the Cartesian coordinate system while dimension refers to the property of a phenomenon by which it occupies space. Matter is anything that has mass and takes up space. So, besides the familiar three–dimensional spatial (physical) world, there are twelve (12) major categories of non-spatial dimensions that include gravity, electromagnetism, strong nuclear interaction, weak nuclear interaction, dark matter, dark energy, mental faculties, senses, emotions, time, life and death. Simply put, the universe is extended in non-physical realms in what can be described as connectedness.

Particle scientists have long attempted to understand the physical world by discovering the smallest divisible matter (atoms, electrons, quarks, etc.) and the surrounding void without success. Today particle scientists are busy searching for gravitons (the supposed carrier particle of the gravitational force) and the Higgs boson (a new kind of elementary particle which acts on others to give them mass). Experiments in the Large Hadron Collider at Cern in 2012 appeared to confirm the existence of Higgs boson. The particle view of nature is a description that works exceedingly well to describe three of the four observed forces of nature. The geometric view of nature works very well for describing gravity at astronomical distance scales. However, efforts to produce a theory of everything have failed because General Relativity and Quantum Mechanics are hard to unify. Therefore, the success of *The Perfect Theory* in part rests on its capacity to reconcile Quantum Mechanics and General Relativity thereby effectively describing all the fundamental interactions and forms of matter.

Albert Einstein was a revolutionary. He suggested, unlike particle scientists, that we should think of space and time as a single entity called space-time. Many technologies including global positioning systems are applications of this observation. However, according to *The Perfect Theory*, it is a limitation to attempt to understand the universe on the basis of where (space) and when (time). Typically the three spatial dimensions (length, width, height), and non-spatial dimensions that include time, are required to describe in a more uniform way the workings of the universe at both the super galactic and subatomic levels. For instance, if we consider the Second World War we want to know where, when, how, who, what and why.

Einstein's general theory of relativity describes the universe on the cosmic scale and applies to stars, galaxies and gravity. Apparently, Einstein's theory fails to comply with the quantum rules that govern the elementary particles (nature at the smallest distance scales). It sounds as if Quantum Mechanics which applies on the subatomic scale of the universe (atoms, subatomic particles and the forces between them) is mutually incompatible with Einstein's theory of relativity. In reality, it is possible to reconcile the two theories to develop a formidable Theory of Everything.

The Perfect Theory combines General Relativity and Quantum Mechanics and applies when it comes to the minuscule and bulk scale of particles and interactions. According to this theory, basic interactions in nature correspond with the macroscopic world (gravity as a bond) as well as the microscopic world of Quantum Mechanics (strong attachment, weak attachment, electromagnetism, and more). The bond of gravity holds together the universe at large, plus the atmosphere, water, and us to the planet Earth. The electromagnetic interaction governs atomic level phenomena, binding electrons to atoms, and atoms to one another to form molecules and compounds. Friction and the explosion of chemical explosive are examples of electromagnetic interactions. The strong nuclear interaction holds the nucleus together. The energy of an atomic bomb (a nuclear bomb) is released as a result of the action of strong nuclear blending. The weak nuclear union is responsible for certain types of outcomes which causes certain forms of radioactive decay. The theory also considers eight (8) additional properties of the universe namely dark matter, dark energy, mental faculties, senses, emotions, time, life and death. In other words, all of the fundamental interactions are different manifestations of a single interaction. The singleness of the physical world (connectedness) represents the unification and simplification of the fundamental interactions.

For many years, scientists thought that the universe was composed almost entirely of ordinary atoms (protons, neutrons and electrons). However, recently emerging evidence indicates that most of the ingredients making up the universe come in forms that we cannot see. Visible matter (atoms) only make up 5 percent of the universe while 23 percent is made up of dark matter which interacts very weakly with ordinary matter, and 72 percent is made of dark energy, which apparently is driving the accelerating expansion of the universe. This realization has prompted a debate on whether matter permeates space (the universe) or space permeates the universe (but remember that nature abhors a vacuum). It is more probable to say that the universe is spatial and non-spatial relations.

Matter [2]can undergo a change of state (i.e. solid, liquid, gas or plasma) with a concurrent energy change. Nonetheless, the third law of thermodynamics considers absolute zero (0 kelvin, or about –273.15 degrees Celsius the temperature at which atoms theoretically stop moving) to be the lower limit for the temperature of any system. Accordingly, no system can reach absolute zero at all through any process that uses a finite number of steps. Each step in the process of lowering an object's temperature to absolute zero can get the temperature a little closer, but you can't get all the way there. Although you can't get down to absolute zero with any known process, you can get close to the near-zero world. Liquid helium will climb entirely out of containers by itself (without friction) at exceedingly very low temperatures (2.17 Kelvin or –269 degrees Celsius). Superfluid Helium (characterized by zero viscosity) never settle into the solid state and can climb walls of the container just like a soft drink can spontaneously

---

[2] The traditional states of matter are solid, liquid and gas but there is also plasma (ions and electrons).

overflow up and out of the drinking straw. Helium's liquidity at very low temperatures undergoes the Bose–Einstein condensation (helium atoms are condensed to the lowest possible energy), in which individual particles overlap until they behave like one big particle. Atoms acting in unison don't behave like individual atoms.

Virtually everything in nature is in motion. Now, the expectation is that every other constituent part of matter will always travel in the straightest possible line if there are no external forces at work. In view of that, without an outside force, two constituent parts of matter travelling along parallel paths will always remain parallel. They will never meet. However, it appears they do meet because the whole universe is connected. In fact, if one could travel in a certain direction on the surface of the earth hoping to reach the end of the universe, s/he could never come up against an impassable barrier or fall over the edge, but eventually come back to where s/he started confirming the oneness of the universe.

Albert Einstein explained this phenomenon in a different way thereby supporting the reality that the whole universe is connected. Einstein noted that things in space have inertia. Objects resist any effort to change their state of rest or motion. Alternatively, objects travel uniformly forward in a straight line unless there is a force that makes them to stop or change. The effects of friction and air resistance, contact or some other source on the earth surface tend to decrease the speed of moving objects to the point of rest. For Einstein, the movement of things in space is influenced by gravity. Nevertheless according to Einstein's theory of General Relativity, gravity is not a force. For him, gravitation is a byproduct of the curvature of space-time governing the motion of inertial objects due to the presence of matter. However, he did not state through which mechanism. Einstein was also aware that most of deep space (the vast stretches of empty area between planets, stars and moons) has abundant (dark) energy. Einstein's space-time curvature is simply a good illustration of connectedness.

Mass, the amount of matter in an object, is relative because matter is extended infinitely. Indeed, to describe the nature of space one has to grapple with twelve integral extensions of the universe including (but not limited to) customary matter. Conspicuously space has an inherent energy that is repulsing to gravity. Everything was smaller and closer together in the past. Indeed, there is a high probability that there was a time when Earth would have been close enough to its planetary neighbors. There must have been some point in time when the universe was half its current size. If you continue to run time backwards, there must have been a time when the universe was an infinitesimally small point. In calculus, this is an adaptation of Zeno's Paradox.

Zeno of Elea presented the paradox that a runner could never actually complete a race. So, in a race, the quickest runner (Achilles) can never overtake the slowest (the tortoise), since the pursuer and the pursued are one and the same. This is because for motion to occur an object must change the position which it occupies. Yet the universe is just about infinite but solitary (connected).

Werner Heisenberg noted that observation affect the behavior of quanta (influences the result of an experiment). Niels Bohr and his colleagues (the Copenhagen interpretation), based on the concept of wave-particle duality and the idea that the observation influences the result of an experiment, suggested rightly that a quantum particle exist in all of its possible states at once. When we make an observation, an object is essentially forced to take one probable state or another. The object may be forced into a different observable state each time it is observed. A wave is spread over a broad region, therefore does not have a specific location. Uncertainty is an integral component of space because at a tiny scale it is impossible to measure both the spatial position of a physical system and its momentum (mass times velocity) with arbitrary accuracy since the particle and the rest of the universe behave as a single entity.

At the heart of the struggle to comprehend the physical world is whether nature is local or nonlocal. Atomists claim nature is local. Einstein also suggests that space is local. Local interactions require a link between subatomic particles. Non-local interactions are where one particle could influence another distant particle in an instantaneous manner (faster-than-light) not via a conventional force field but simply because they touched one another sometime in the distant past. Quantum theorists demonstrate that quantum reality is non-local.

According to *The Perfect Theory*, to render a more complete account of the physical world we have to admit the undivided wholeness inherent in nature. Therefore, the nature of the universe is both local and nonlocal.

The Perfect Theory assumes that the universe is composed of **spatial** and **non-spatial extensions**. The theory implies that besides the familiar three–dimensional physical phenomena, there are twelve (12) major categories of non-physical phenomena: gravity, electromagnetism, strong nuclear interaction, weak nuclear interaction, dark matter, dark energy, mental faculties, senses, emotions, time, life and death. This information is summarized in the diagram on the next page:

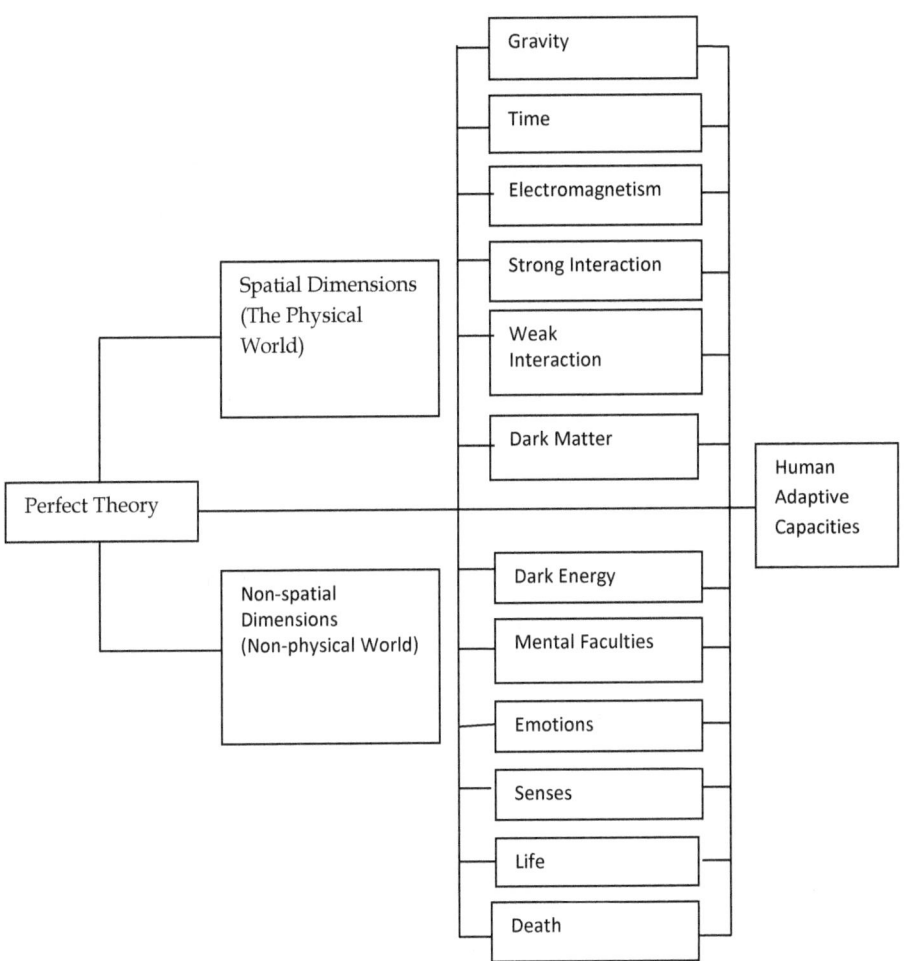

Figure 3: An Illustration of Connectedness

Spatial Dimensions

The fundamental nature of the universe remains an interesting topic for discussion in science. Since the prehistoric times, humans have been curious to comprehend nature. Today scientists have one theory (General Relativity) to describe the behavior of large masses such as planets, stars and galaxies and another theory (Quantum Mechanics) to explain the behavior of the various particles that make up atoms and the forces that work in the subatomic territory. Immanuel Kant (1724-1804) in his work *Critique of Pure Reason* noted that the phenomenal world (phenomena) should be the subject for scientific study because the *noumenal* (true) world (the thing-in-itself) is not knowable. According to Kant, the intellect plays a central role in influencing the way that the world is experienced. In Kantian terms, sensory and mental representations are mere phenomena. We explain the world using mental concepts such as space and time. However, space and time are not substances but a priori particulars that enable us to comprehend sense experience. Substances are the basic things out of which the world is composed. Enchantingly, the nature of 95 percent of physical reality is literally obscure comprising of dark matter, dark energy, etc. Ordinary (visible) matter that incorporates material objects such as stars and galaxies comprises only a small fraction (5%) of the constituents of the cosmos. The rest of the universe is far from empty, however. Together, dark matter and dark energy make up the bulk of the universe.

Spiritual and early philosophical traditions in almost all parts of the world have long held that the universe is boundless and linked in one piece. Surely, *The Perfect Theory* holds that the universe (nature) is an organic whole, a whole in which all parts are connected in conformity with many early prepositions. Nature is not a total vacuum. The universe is uninterrupted and bursting with activity/interactions.

The universe is made up of spatial (what is commonly referred to as the physical world) and non-spatial extensions. Spatial dimensions represent the physical world (noumenal or the thing-in-itself). The physical world cannot be comprehended directly. Physical space is grasped through twelve (12) non-spatial dimensions (phenomena): gravity, electromagnetic force, strong nuclear interaction, weak nuclear interaction, dark matter, dark energy, mental faculties, senses, emotions, time, life and death. The interconnections and interactions that are characteristic of nature give rise to connectedness (a universe that is largely a continuous whole and that is not demarcated).

Connectedness totally explains the phenomenal world. According to *The Perfect Theory*, the universe at its smallest and largest distances exist as undemarcated whole. The principle of connectedness is the reason as to why quantum theorists have noted uncertainty relations as summarized by Heisenberg's uncertainty principle. Attempts to measure a particle's position must randomly change its velocity and speed. The universe is linked giving rise to an unbroken whole such that a quantum physical system doesn't exist in one state or another but in all of its possible configurations simultaneously but is only forced to take the state that we sense (observe or measure). So, the unanimity of the universe is the reason as to why the situation and velocity of small particles is influenced by simply observing the quanta and is also the reason as why telephonic conversations (and internet communications) are immediate. This is also the reason as to why the magnetic compass will right away predict the direction of the earth's magnetic field. Observably, two magnets "feel" each other's presence and attract or repel without touching whenever wherever.

Connectedness informs our understanding of nature. Connectedness embraces both Quantum Mechanics and relativity because actually the two theories do not offer different interpretations of the universe. Connectedness is the reason as to why the phenomenon of entanglement envisages correlations between remote events. Indeed Niels Bohr stressed the undivided wholeness inherent in the quantum mechanical description of nature. Such that changes made in any part of the physical system result in overall changes to the whole system. Connectedness is also the reason as to why Einstein showed demonstrated that on the large-scale space and time are inextricably linked – space-time continuum.

The Perfect Theory holds that an object is directly influenced both by its immediate surroundings (locality) and distant objects/events (non-locality). This view ultimately rests upon the assumption that the material universe has non-material extensions. Therefore, even when material objects are separated by large distances (potentially even trillions of light years) they are actually potentially connected in an immediate and instantaneous manner. Nonlocality occurs due to the phenomenon of connectedness. Seemingly separate parts of the universe communicate instantaneously violating Einstein's suggestion that the speed of light is the maximum velocity for anything in the universe.

Causality is integral in science and in our everyday life. Causality is such that a cause always precedes its effect. Simply put, events in the past contribute to the occurrence of events in the present and what happens in the future is caused by events in the present. These limitations are based on the supposition that causal influences cannot travel faster than the speed of light and/or backward time travel. Causality rules out systems that change from one state to another without any apparent physical trigger. It is also rules out the possibility of conceiving situations in which a single event can be both a cause and an effect of another one. Nevertheless, results of tests of Bell's theorem appear to demonstrate that some quantum effects travel faster than light. Cleve Backster in *Evidence of a Primary Perception in Plant Life* (1968) established that plants react to trauma in their local environment. The Backster Effect showed plants can perceive and measurably respond to intentional human thought and actions through emotions. Plant perception does not deviate from our everyday view of reality. The universe as a whole reacts instantaneously but the effects are observed gradually over time. The first significant activity was the dawn of the universe giving birth to science. Science/evolution is the practical and intellectual adaptive activity capacity to preserve life (endlessly in the long term).

The EPR paradox correctly demonstrated that the phenomenon of entanglement violates principles of local causality. Causality is pegged on Einstein's proposition that particles can never move faster than the speed of light through a vacuum. To rewrite this law (if the light-speed barrier can be overcome) envisions the possibility of time travel thereby undermining the laws of causality. However, Einstein was mistaken to imagine that nature is local. The physical world is both local and non-local. That being the case, causality is a consequent process not an event. Causality is the outcome of inestimable events. Causality is merely the customary sequence of apparent causes. Seemingly the entire universe is organized in such a way as to anticipate the past and the future.

The nature of space is both local and non-local. Events and processes in the universe that culminated in the beginning of the universe, evolution, etc. demonstrates the connectedness of physical world in the immediate vicinity (local) and large distances (non-local/interaction without touching). All events and processes that take place in nature, irrespective of the distances that separate them, are linked, uninterrupted and related in an intimate and instantaneous manner. This is why in the process of interaction pairs particles effectively lose their individuality and in many ways behave as a single entity refuting Einstein's speed of light theorem. However, the effect(s) (causality) is gradual and ongoing. Some notable effects include the origin of the universe and the subsequent evolutionary process, slavery, colonialism, wars, trade, marriage, ethnicity, the state, urban spaces, etc.

The Big Bang framework is now the standard cosmology describing the early development of the universe. According to the Big Bang model, the universe came into existence with one enormous explosion of energy and light 13.7 billion years ago. Milliseconds after its initial primeval explosion from the early hot and dense state (a singularity), the universe cooled, coalesced and compressed sufficiently under gravity to allow some of the energy (clouds of gases) to be turned into various subatomic particles, including protons, neutrons, and electrons. The intense energy released in the cosmic big bang provided the source of all the matter in the universe. From the beginning the universe is separate but remains connected by an instantaneous wholeness. The connectedness of the universe depends on twelve (12) non-spatial extensions that include gravity, dark matter and dark energy.

Quantum theory though manifestly non-local, acceptably suggests that whatever we can observe (quantum facts) are always local. Nonetheless, quantum theorists are incorrect to postulate that the possibility of observing a non-local effect is remote (or rather non-existent). This is because local and non-local effects are what we call experience (trial and error). We observe (or rather experience) them every day. According to *The Perfect Theory*, nature at its smallest and largest distances is characterized by uncertain relations (randomness) because local and non-local effects are gradual and ongoing. The current state of the universe that include diversity of life, natural phenomenon such as earthquakes, rainfall, tides, etc., experienced life including marriage, prostitution, war, diplomacy etc., and the advance of science are all gradual and ongoing effects of the birth of the universe. The discovery of agriculture and sedentism, renaissance, reformation, enlightenment, major wars (WW1, WW2 and the cold war) and colonialism, for instance, are consequent events that continue to influence worldviews in the present and possibly in the future. Space is not empty. This reality explains natural events and processes (i.e. rainfall, earthquakes, volcanic eruptions, tides, gravity, etc.) and human events and processes (i.e. schooling, marriage, science, religion, crime, war, poverty, underdevelopment, urbanization etc.). In other words, local and non-local effects explain experience (evolution). Experience is a process not an event. Experience is connectedness.

Since the beginning the universe has been evolving. Evolutionary processes have given rise to many natural processes, repeated speciation, divergence of life at every level of biological organization, and extinction. Adaptation allows organisms to be better suited to their habitat(s) for an organism's survival and reproduction. Adaptation occurs through the gradual modification of existing structures. Darwinian evolution is a slow, gradual process. Today proponents of evolution are largely divided into two camps. There are those who suggest that natural selection is the principle guiding evolution. For pre-selectionists, natural selection works to create a population that is highly suited to its environment and that can adapt to changes. Also, the useful traits that give individuals a survival advantage and the chance of reproducing are subject to sexual selection. The other camp opines that natural selection does not explain the full complexity of evolution. Without a doubt, some facets of Darwinian struggle for survival are mistaken.

First, Charles Darwin's main preoccupation was biological evolution (descent with modification). Accordingly, all living species are the modified descendents of earlier species. Volcanoes erupt, trees lose their leaves, mountain ranges rise and erode, a river may change its course (a geographical change), but they aren't examples of biological evolution because they don't involve descent through genetic inheritance. Biological evolution has produced the diversity of life on Earth. Nonetheless, evolution is not restricted to living things. Non-living things also evolve via the human activities or other natural processes.

Second, the assumption that evolution is exclusively driven by a process of natural and sexual selection does not rhyme with empirical evidence. Darwin argued that all individuals struggle to survive on limited resources, but some have small, heritable differences that give them a greater chance of surviving or reproducing, than individuals lacking these beneficial traits. Such individuals have a higher evolutionary fitness, and the useful traits they possess become more common in the population because more of their offspring survive. Eventually these advantageous traits become the norm. Conversely, harmful traits are quickly eradicated as individuals that possess them are less likely to reproduce. Natural selection therefore works to create a population that is highly suited to its environment, and that can adapt to changes. Therefore, natural and sexual selection are simply some of the mechanisms the universe exploit to preserve life.

Third, the conjecture that death is inevitable and a necessary component of the evolution process, is somehow awkward, though factual. Evolution theory postulates that individuals in a population and species struggle amongst themselves to survive, and most become extinct over time. Sometimes species die out in mass extinctions such as the one that caused the demise of the dinosaurs. Recently, some scientists have suggested that we are the throes of another mass extinction as a result of human overexploitation of habitats. It is highly improbable to envisage that life as we know it is the product of a succession of accidents. It appears that, contra to

Darwinian postulations, the universe is structured to preserve life endlessly (if and when possible).

Evolution does not only happen by way of natural selection. Therefore, a more complete and accurate understanding of Darwinian thought requires the appreciation of other mechanisms of evolution that include genetic drift, gene flow and mutation.

The Perfect Theory is the closest humanity has ever come to having a complete unified description of nature. The central contention is that the universe is best grasped in terms of the frame of reference [3]and non-spatial dimensions which are intrinsic features of the universe. According to *The Perfect Theory*, electromagnetism, the weak interaction, the strong interaction, gravity, dark matter, dark energy, mental faculties, senses, emotions, time, life and death are properties of space itself.

This unified model can account for phenomena that occur at very small distances in the realm of atoms and subatomic particles, that is, electromagnetism (which holds atoms and molecules together), the strong interaction (which holds protons and neutrons together), and the weak interaction (which causes certain forms of radioactive decay). The model also captures the effects of gravity around large masses. The theory explains the formation of the universe, why the moon orbit around the Earth, why humans experience a wide range of emotions and thoughts, why life-forms respond to physical and chemical stimuli and account for many other phenomena in the universe. The theory also offers a more reliable account

---

[3] The frame of reference refers to position, and orientation or direction

on why it is not possible to measure the position and direction of an object.

Greek philosophers Leucippus, Democritus, and Epicurus (5th and 4th century BC) suggested that there may be other inhabited worlds. Indeed, according to some variants of quantum gravity theory, there exists the possibility that our universe is just one among many multiple universes. This thinking suggests that we exist in an inestimable number of universes constituting the multiverse. In fact, a number of religions talk of an afterlife existence in realms, such as heavens and hells, which are apparently different from the observable universe. Some quantum theorists suggest that possibility of various unobservable universes (and alternate histories) within a greater multiverse. The splitting of the universe arises from the interaction between states and the environment producing entanglement. However, consistent with *The Perfect Theory*, it is a contradiction to imagine additional spatial dimensions.

The principal anomaly with the parallel universe theory is the suggestion that multiple universes exist as distinct entities without any interaction. According to connectedness, there is only one thing (physical world/frame of reference) and what seems to be a plurality is merely a series of different aspects of this one thing. Therefore, it is more probable to suggest that the phenomena that we cannot account for arises due to the twelve (12) non-spatial dimensions. Accurate awareness of the unity and mutual interrelation of all events and processes makes us aware they represent different manifestations of the same ultimate reality – the solitary physical world. So, probably there is only one universe, our universe. Archimedes of Syracuse (287 BCE – 212 BCE) put it better. If given a fulcrum and a place to stand, it is possible to move the world. Every point can be regarded as the center of the universe.

The physical world is linked and interacts. These interactions include attractions and repulsions, decay, and annihilation. For example, the North Pole, interact/exist with the South Pole instantaneously. The glue that holds the universe together on the large-scale and small-scale is connectedness - the mutual attraction of everything in the universe for everything else. When the universe was young, it was nearly smooth and featureless. As it grew older and developed, it became organized into planets (including the Earth!) orbiting around the Sun, galaxies and clusters of galaxies. The universe (the thing-in-itself) is not directly knowable. A proper understanding of the universe requires the comprehension of its twelve (12) constituent non-physical properties.

Non-Spatial Dimensions

Non-spatial dimensions are human conceptual structures, not real existing things separate from nature itself. Gravity, electromagnetism, strong interaction, weak interaction, dark matter, dark energy, mental faculties, senses, emotions, time, life and death are properties of nature. Non-spatial dimensions envisage non-material existence in physical space. Non-spatial dimensions are simply bonds of connectedness.

Strong Interaction

The strong interaction (also known as the strong nuclear force or nucleon interaction) is the binding mechanism of the atomic nucleus (combine protons and neutrons into atomic nuclei). The present-day understanding of strong force is described by Quantum Chromo-Dynamics (QCD), a component of the Standard Model of particle physics. The strong interaction is based on the understanding that like charges repel (+ +, or - -), and unlike charges attract (+ -). The strong interaction holds together the subatomic particles of the nucleus (protons that carry a positive charge and neutrons that carry no charge). These particles are collectively called nucleons. The strong attraction only operates upon elementary quark and gluon particles. It has a short range; the import being that particles must be extremely close before its effects are felt and it rapidly diminishes in strength with increasing distance. The strong interaction is what is liberated during nuclear reactions, of the sort that take place in the Sun, nuclear power plants, and nuclear bombs.

Electromagnetism

Electricity and magnetism were long thought to be separate forces until the unification work of Michael Faraday and James Clerk Maxwell. Electricity and magnetism are two aspects of electromagnetism. Electromagnetism broadly refers to the properties of electric and magnetic fields. Electromagnetism, at the subatomic level, is the interaction that causes the attraction and repulsion of electrically charged particles. It has both attractive and repulsive properties that can combine or cancel each other out. It comes in two charges: positive and negative. Two positive or two negative things will repel each other, but one positive and one negative attract each other. This is the principle that keeps atoms together: the positively charged nucleus and the negatively charged electrons attract each other. The messenger particle of electromagnetism is the photon, a massless particle that logically travels at the speed of light (since light is a manifestation of electromagnetism).

Electromagnetism is the interaction responsible for practically almost all the phenomena encountered in daily life, with the exception of gravity. Ordinary matter takes its form as a result of intermolecular forces between individual molecules in matter. Electrons are bound by electromagnetic wave mechanics into orbitals around atomic nuclei to form atoms, which are the building blocks of molecules. If one rubs a comb across the clothes one is wearing to give it static electricity then hold it over a piece of paper on a desk, the piece of paper lifts off the desk. This is an everyday example of electromagnetism. The quandary is that things seem to interact without touching! How do two magnets "feel" each other's presence and attract or repel in view of that? How does the sun attract the earth? When electrically charged particles, such as electrons, are put into motion, they create a magnetic field. When these particles are made to oscillate, they create electromagnetic radiation. This can include radio waves, visible light, or x-rays, depending on the frequency of the oscillation.

The Earth has a magnetic field, as can be seen by using a magnetic compass. The magnetic field is largely that of a dipole, one North Pole and one South Pole. There are currently two poles observed on the surface of the Earth, one in the Northern hemisphere and one in the Southern hemisphere. In the early 20th century geologists first noticed that some volcanic rocks were magnetized opposite to the direction of the local Earth's field. Later, when Earth's magnetic field was better understood, it emerged that the Earth's field might have reversed in the remote past. A geomagnetic reversal (of polarity) is a change in the Earth's magnetic field such that the positions of magnetic north and magnetic south are interchanged. The North Pole is transformed into a South Pole and the South Pole becomes a North Pole. The Earth's field has alternated between periods of normal polarity, in which the direction of the field was the same as the present direction, and reverse polarity, in which the field was the opposite. Reversals are not predictable and are certainly not periodic in nature. Most reversals are estimated to take between 1,000 and 10,000 years. Not long after the first geomagnetic polarity time scales were produced, scientists began exploring the possibility that reversals could be linked to extinctions.

Weak Interaction

Weak interaction (often called the weak nuclear force) is responsible for the radioactive decay of subatomic particles (radioactivity). The weak interaction operates only on the extremely short distance scales found in an atomic nucleus. Weak interaction represents connectedness of physical world in the immediate vicinity.

Gravity

There is gravity everywhere. Gravity is most commonly experienced as the agent that gives weight to objects with mass and causes them to fall to the ground when dropped. Gravity gives shape to the orbits of the planets, the solar system, and even galaxies. Gravity from the Sun reaches throughout the solar system and beyond, keeping the planets in their orbits. Gravity from Earth keeps the Moon and human-made satellites in orbit. Things in space have inertia. That is, they travel in a straight line unless there is a force that makes them stop or change. The movement of things in space is influenced by gravity. While some objects in space travel in irregular paths, most (especially our near neighbors in space) tend to travel in orbits around the Sun or around planets.

Isaac Newton described gravity as a natural predictable force (a function of both mass and distance) that attracts all objects to all other objects. For him, physical bodies appear to attract each other with a force proportional to their masses. Gravity causes any two objects with mass in the universe to be drawn to one another with a force that is directly proportional to the product of their masses and inversely proportional to the square of the distance between them. Mass is how we measure the amount of matter in something. The more massive something is, the more of a gravitational pull it exerts. In addition to depending on the amount of mass, gravity also depends on how far you are from something. Gravity decreases with distance. Objects far away from a massive object such as a planet or star feel less gravity.

Albert Einstein did not believe gravity was a force at all. He said gravity is a distortion in the shape (the curvature) of space-time. Gravity can bend space-time through a yet-to-be discovered mechanism. These two theories are the most common and widely held (if somewhat incomplete) explanations of gravity. *The Perfect Theory* holds that gravity is because the universe is connected.

If there are no external interventions at work, an object will always travel in the straightest possible line. Accordingly, without an external intrusion, two objects travelling along parallel paths will never meet. However, in real life particles that start off on parallel paths sometimes end up colliding. Newton's theory says this can occur because of gravity, a force attracting those objects to one another or to a single, third object. Einstein also says this occurs due to gravity but for him gravity is a curve in space-time not a force. According to Einstein, those objects are still travelling along the straightest possible line, but due to a distortion in space-time, the straightest possible line is now along a spherical path. So two objects that were moving along a flat plane are now moving along a spherical plane. And two straight paths along that sphere end in a single point. Recent theories express the phenomenon of gravitation in terms of particles and waves. Particles called gravitons cause objects to be attracted to one another. Gravitons have never actually been observed, though. And neither have gravitational waves (or radiation), which supposedly are generated when an object is accelerated by an external force. All these theories seem to agree that the universe is linked. *The Perfect Theory* suggests that two objects travelling along parallel paths will meet because the universe is connected.

Gravity assisted in forming the universe. Gravitation causes dispersed matter to coalesce, and coalesced matter to remain intact, thus accounting for the existence of planets, stars, galaxies and most of the macroscopic objects in the universe. Gravity keeps the Earth and the other planets safely in their orbits around the Sun; keeps the Moon in its orbit around the Earth; is responsible for the formation of tides; and for various other phenomena observed on Earth and throughout the universe.

Gravity is an integral property of the physical world. We study properties (phenomena) of the physical world to grasp the universe. From early times debates have raged over the reality of motion and rest (acceleration). Speed (a scalar quantity) refers to the rate at which an object covers distances (how fast an object is moving) relative to other bodies. An object with no movement at all has a zero speed. A fast-moving object has a high speed and covers a relatively large distance in a short amount of time. A slow-moving object covers a relatively small amount of distance in the same amount of time. Velocity (speed with a direction) is the rate at which an object changes its position. Velocity gives both how fast and in what direction the object is moving. To have a constant velocity, an object must have a constant speed in a constant direction (motion in a straight path). The object's path does not curve. An object is said to be undergoing an acceleration (changing its velocity) if there is a change in speed, direction, or both. Nonetheless, the equivalence/elevator principle and the Heraclitean and Parmenidean dispute over rest and motion indicate that gravity and acceleration are the same.

Einstein's equivalence principle states that a uniform acceleration is equivalent to a uniform gravitational field like the one on Earth. Suppose an elevator is located in space and is accelerated upwards. The passenger in the cabin feels a downward pull that is equivalent to the pull of the gravitational field on Earth. Suppose the elevator is located on Earth and is in the state of free fall. The passenger in the cabin feels no gravity, because the gravitational field of the Earth is cancelled by the opposite acceleration of the elevator. In both cases, the person cannot tell the difference between the pull of acceleration and gravity, or respectively the weightlessness felt in space and on Earth. According to the elevator principle, a falling object does not "feel" any gravitational force, while an object being accelerated does. Einstein equated gravitational mass with inertial mass.

Zeno of Elea (490 BCE – 430 BCE) in an argument over an arrow in flight logically observed that nothing can be in two places at once (i.e. for motion to occur an object must change the position which it occupies). He, therefore, reasoned that an arrow is only in one place during any given instance of its flight. But if it is in only one place, it must be at rest. The arrow must then be at rest at every moment of its flight. If the arrow is motionless at every instant, then motion is impossible.

Around 350 BCE, Aristotle in *Physics* paraphrased Zeno and put forward the view that mechanical objects prefer the state of rest if there is no external force sustaining motion. This proposition is reminiscent of the classical controversy between the Greek schools of Heraclitus and Parmenides. Parmenides held that all is one and that motion is an illusion, while

Heraclitus stated just the opposite, namely that motionlessness is an illusion and that everything is always in a permanent state of motion and change.

According to *The Perfect Theory* opposites cohere; gravity and acceleration are the same. In view of that, the physical world is both Heraclitean and Parmenidean. The motion of all objects is relative to each other (because the universe is uninterrupted whole), and it is really a matter of convention to define one reference frame as being at rest while the other is considered accelerated. An object that appears to be at rest in a designated reference frame is in motion in another one. Similarly, an object that appears to be accelerated in a designated reference frame is at rest in another one. Differently put, acceleration is equivalent to gravitation.

The universe is linked. Acceleration and gravitation are the same because the universe is connected. The physical world is one. Reality is space (spatial dimensions) and its properties (non spatial dimensions). Nature is both local and non-local and as such there is some kind of instant (faster than light) communication between events or particles in the physical world at both the smallest and largest distances. Gravity (and acceleration) is not a material thing but one of the properties of the physical world. In this case, the cabin and the passenger cannot be thought of as having any clear-cut existence in terms of either extent or position in space. This is because all objects (living and non-living) are mutually dependent and undividable parts of the entire Universe. For that reason Werner Heisenberg's principle of indeterminacy correctly demonstrates that it is not possible to exactly determine pairs of physical properties of a particle such as the velocity and location.

Gravity is dominant on macroscopic scales because of its enormous range and constant attractive nature. Gravity binds all objects in the universe. Gravity is simply connectedness writ large. The range of gravity is unlimited - every object in the universe exerts a gravitational effect on everything else relative to the mass of two bodies and the distance between them.

Dark Energy

Albert Einstein observed that space is not empty and it is possible for more space to come into existence. He also suggested that "empty space" can possess its own energy. Dark energy is present throughout the universe and fills the vast reaches of space. In fact, it makes up nearly three quarters of the universe. It is the force that works against gravity. From 1998, the Hubble Space Telescope (HST) observations of very distant supernovae showed that a long time ago the universe was actually expanding more slowly than it is today. So the expansion of the universe has not been slowing due to gravity as everyone thought but it has been accelerating. Dark energy is responsible for the accelerating expansion of the universe at an ever-increasing rate since the birth of the universe. According to *The Perfect Theory*, dark energy is a non-spatial property of nature.

Dark Matter

Dark matter is a type of matter that cannot be seen directly with telescopes and evidently neither emits nor absorbs light or other electromagnetic radiation at any significant level but is hypothesized to account for a large part of the total mass in the universe. This invisible stuff is thought to outnumber regular matter in the universe by a factor of about six to one. Dark matter consists of the unseen particles that keeps the universe together and explains how the cohesion of the stars, galaxies and more is even possible. Dark matter doesn't emit or reflect light, but scientists can estimate where it is based on its gravitational effects on surrounding visible matter. High concentrations of matter bend light passing near them from objects further away. Dark matter is not antimatter because we do not see the unique gamma rays that are produced when antimatter annihilates with matter. Craig Freudenrich established that dark matter is located by its gravitational effects on its surroundings and the detectable X-rays it emits. Gravitational lensing enable scientists to observe the way dark matter's gravitational pull bends and distorts light from distant galaxies. Some theoretical physicists suggest that dark matter is made of WIMPS (weakly interacting massive particles).

Senses

The sensory faculties (senses) are physiological capacities of organisms through which the external world is grasped. Senses provide data for perception. The senses include taste, smell, sight, hearing, touch, pain, balance, body position, temperature differences among others. Humans possess the ability to detect other stimuli beyond those governed by the traditional senses. The nervous system has a specific sensory system or organ, dedicated to each sense. People connect with the physical world by touching, tasting, smelling, seeing, and hearing things around them. Non-human life forms may possess senses that are absent in humans and are, therefore, able to sense the world in a way that humans cannot. Some animal species possess the capacity to sense changes in electrical and magnetic fields, detect water pressure and currents, detect polarized light, determine orientation and detect the direction one is facing based on the Earth's magnetic field. On the whole, senses are the faculties by which stimuli from outside or inside the body are received and felt to facilitate oneness with the rest of the universe.

Emotions

Emotion encompasses a wide range of psychological phenomena among human beings and other animals in the process of interacting with biochemical (internal) and environmental (external) influences. Reason is also often in opposition to emotions. Emotion is associated with mood, temperament, personality, disposition, and motivation. Basic emotions include love, happiness, compassion, empathy, nurturance, altruism, fear, sadness, mourning, anger, aggression and disgust.

Scientists have long struggled to explain why most creatures reproduce sexually when they could just clone themselves. Sexual reproduction, as a process, facilitates the combining of genes (specialized cells/gametes) to form offspring that inherit traits from each parent. Sexual reproduction facilitates the change in the frequency of alleles (new traits) in populations over time. Biological sex (i.e. the state of being male or female) is an important aspect of life because the choice of a mate is a random event. Sex distinctions (sex as a biological category) influence the formation of gender identity (social distinctions based on sex). Sexual reproduction evolved to encourage oneness. Love is a virtuous emotional state of oneness characterized by a strong affection, personal attachment and kindness towards one's self, other humans, animals etc. Love may describe passionate desire and intimacy (romantic love), sexual love (eros), familial love (emotional closeness), friendship (platonic love), devotion (religious love) or a profound oneness encompassing all of those feelings. Love in its various forms acts as a major facilitator of interpersonal relationships and connectedness to the rest of the universe. When understood this way, love is part of the survival instinct that facilitates the continuation of life.

Love is often driven by either attachment or sexual attraction. For instance, the family is partly meant to provide emotional ties but is also a mating arrangement (exclusiveness). The family is a mechanism that every member of society will find someone to mate with (inclusiveness). The most noticeable aspect of the family is the incest taboo and the consequent identification of family members and relatives. The incest taboo is universal. It is suggested that because inbreeding leads to detrimental outcomes humans have learned to avoid sexual partners with whom one is likely to share genes. There is also the view that the incest prohibition is a deliberate cultural attempt to discourage endogamy in preference for exogamy for purposes of forging valuable alliances among otherwise unrelated households. The social solidarity resulting from such arrangements informs the construction, reinforcement and perpetuation of socio-political institutions that include lineage and descent groups, clans, ethnic groups, states, religions, races and civilizations.

Over eons and many generations it been noted that there exist a strong relationship between music, mind (sensory perception/intellect) and emotions especially love. Music's capacity to express emotion reveals a lot to us about the physical world. Indeed, music can moderate our lonesomeness or provoke our passions. The philosopher Plato suggests in the *Republic* that music has a direct effect on the heart requiring close regulation by the state. Arthur Schopenhauer (1788 CE – 1860 CE), the foremost pessimist, in *The World as Will and Representation* (1818) saw in music humankind's illusionary and self-defeating attempt to escape destiny. There is also a strong connection between music and mathematics. Randomness which is a fundamental feature of nature also informs music. Music is a random human composition/performance that makes sense by alternating sound and silence. Music is often ordered and pleasant to listen to through the alternation of pitch, melody, harmony, rhythm, beat/pulse, tempo, meter, articulation, allocation of voices, dynamics, structure/form, timbre and texture. Music is composed and performed for many purposes, ranging from aesthetic pleasure, religious/ceremonial purposes, or as an entertainment product for the marketplace. Music is experienced by individuals in a range of social settings ranging from being alone to being in a large group. The border between music and noise is determined by occasion, relevance, etc. Actually, in music humans are able to unify the experienced life (mortal life) with the desired life (infinity). Music (like literature, art, scenic beauty and heightened emotion particularly love) is a form of connectedness.

Emotions as part and parcel of the process of closeness find expression in our everyday life in the form of the handshake and mourning. The handshake is a brief habit in which two people grasp one of each other's right hands often accompanied by a brief up and down movement of the grasped hands. At times after the handshake the palm is placed unto the heart. Now and then a kiss may be used instead of a handshake. The handshake is frequently done upon meeting, greeting, parting, offering congratulations, expressing gratitude, completing an agreement or as a sign of good sportsmanship. The handshake is not an event but a process. The origin of the handshake is debatable. For some, the practice originated as a gesture of peace to demonstrate that the hand holds no weapon. For others, the rite serves to express commitment, trust, stability, and parity. Either way, the handshake is a physical and emotional activity to articulate connectedness. Mourning is the public display of emotion of unhappiness at the death of someone.

Points of emotional (and even physical) connectedness include the marketplace, centres of learning and research, sporting arenas, public resting places, offices, waiting lounges, prayer centres, urban/city life, etc. The human society is structured in such a way as to encourage social mingling especially courtship, schooling, parties and celebrations, music festivals, sports events, interactions on social media, among many others.

The escalation of urban spaces (and the consequent urban life) is also a pointer to connectedness. The current population of the world is approximately 7.0 billion people out of which about 3 billion live in cities. Estimates indicate that by 2030, population in cities is expected to hit 4.1 billion. The city (urbanization) is a relatively recent form of social organization. Urbanization has been closely connected with the application of more and more inanimate sources of energy to enhance human productivity to produce surpluses (industrialization). The concentration of people caused many villages to grow into towns and cities. This also enabled specialization of labor and crafts, and development of many trades. Today villages have been eclipsed in importance as units of human society and settlement. Larger and larger proportions of people live in cities. Rural areas have experienced notable population loss (out-migration) as a result. The movement of people from remote locations to overcrowded urban spaces (urbanization) is simply the quest to turn the globe into a world village (connectedness).

Mental Faculties

Mental faculties are natural cognitive or perceptual powers of the mind. There are six major components of mental faculty: reason, imagination, perception, intuition, memory and will. Reason (deduction or induction) enables human beings to draw conclusions from ideas, situations, or premises. Imagination is the capacity to construct images, sensations, or concepts at a time when not being perceived by physical senses. Imagination is largely affected by reason and vice versa. Perception (worldview) is the process by which people gain an understanding of the information we attain through the physical senses. Perceptions differ. Intuition is a priori knowledge or extra-sensory perception. Memory is the human capacity to retain and recall information. Willpower (or will) is what facilitates action. Mental qualities assist in the understanding of natural phenomena, human activity and social patterns. Natural phenomena are non-artificial event in the physical sense. Types of natural phenomena include: geological phenomena (volcanic activity and earthquakes); meteorological phenomena (rain, hurricanes, thunderstorms, and tornadoes); and oceanographic phenomena (tsunamis, ocean currents and breaking waves). Human behavior and social environments are beliefs and desires as evidenced in socio-political institutions (i.e. family, lineage, clan, ethnic diversity, religion(s), race, civilizations) and culture (i.e. symbols, myths and rituals).

Life

Science revolves around life (and the struggle to escape from death). A succinct all-inclusive definition of life has been difficult in part because life is a process, not a wholesome substance. It is this predicament that made the physicist Erwin Schrödinger, who discovered the wave function, in his *What Is Life? Mind and Matter* (1944) to ask: "how can the events in space and time which take place within the spatial boundary of a living organism be accounted for by physics and chemistry?" Schrödinger pioneered the idea of an "aperiodic crystal" that was picked up biologist James D. Watson and the physicists Francis Crick, co-inventors of the double helix molecular structure of deoxyribonucleic acid (DNA), the chemical substance that genes are made of. For lack of a proper widely accepted definition of life, scientists highlight fundamental properties of life from all levels of complexity:

i. Life forms contain genetic material (DNA) and can replicate themselves.

ii. Life forms are composed of one (single-celled creature) or more (many-celled organisms) cells. Cells are typically considered the basic unit of life as they contain DNA and can interact with their environment.

iii. Living things respond to their environment.

iv. Living things are highly organized systems. Even the simplest life form is tremendously complex and is capable of effortlessly performing complex transformations and operations.

v. Living things harvest and use energy to maintain themselves. Plants, for instance, take in energy by absorbing sun light. This energy is converted to chemical energy by a process called photosynthesis. Humans take advantage of the products of photosynthesis as the ultimate source of our food.

vi. Living things exhibit metabolism; are capable of transforming carbon-based and other compounds into useful energy using chemical reactions. Food digestion breaks food down into simple compounds that serve as a source of energy to power the building of other compounds.

vii. Living things are the product of evolution. Charles Darwin (1859) proposed that complex life evolved from simpler beginnings, exploiting the materials around it for its own survival. Populations of living things evolve via natural selection in response to changes in the environment. Living organisms have many structures and behaviors that enable the organism to survive in its environment. Adaptation involves acquiring advantageous traits in a way that the species could change forms and functions gradually over long periods of time.

All living organisms share most of these characteristics in common. However, certain animals such as brine shrimp have eggs or other resistant stages in their life cycle that can dry out, show no signs of life, and remain viable for many years. A desert plant called *Lithrops* (literally stone) are so slow growing and perfectly blend in with the desert that they could easily be mistaken for rocks. Do viruses also belong to life? Viruses are simple, having often only protein coat and one or a few molecules of nucleic acids. They are not considered living by most biologists because they are not made of cells, do not have independent metabolism and do not take in energy. For reproduction they require the higher organisms in which they reside. Nonetheless, some biologists consider them alive because they possess individuality (have their own genes or DNA) that enables them to target specific host species.

Viruses are simple creatures. So simple they don't eat, drink or breed — they're just a bunch of genes dressed up in a protein coat. They get by invading living cells (like ours), hacking their software (the DNA) and turning them into their own personal virus-making factories.

Earth is the only planet known to harbor life. Life on Earth occurs in a bewildering array of forms namely, humans, animals, plants and microorganisms, which together compose Earth's biosphere or Gaia. The idea that the Earth is alive is probably as old as humankind. James Hutton (1785) hypothesized the idea of a living earth; he stated that the Earth was a super-organism. Later, James Lovelock (1919-) in the Gaia theory/principle suggested that the life on Earth functions as a single organism which defines and maintains environmental conditions necessary for its survival. The Gaia hypothesis proposes that all organisms and their inorganic surroundings on Earth are closely integrated to form a single self-regulating complex system that maintains life on the planet. The self-regulating system link the biosphere, the atmosphere, the hydrosphere, cryosphere, lithosphere and the pedosphere, tightly coupled as an evolving system. The theory maintains that this system as a whole, called Gaia, seeks a physical and chemical environment optimal for contemporary life.

For many years it is hypothesized that life might exist elsewhere in the universe in the form of extraterrestrial life. Scientists have found evidence of Earth-like planets orbiting distant stars. Whether any of these planets will harbor life is an open question. Currently scientists are engaged in ongoing efforts to discover sign of life on Mars and other planets. It is even argued that planets, stars and other terrestrial objects are living things.

During antiquity, Anaximander, an ancient Greek philosopher, and Eastern mysticism particularly Hinduism, Buddhism, Taoism and Zen indirectly suggested that life

71

develops from non-life. This they did through the concept of origination and the notion that all things descended from one central guiding principle. Aristotle implied a transition from non-living to the living through the observation that all things constantly desire to move from the lower to the higher realm finally becoming the divine. Darwin's theory of evolution explains the surfacing of human beings and other forms of life. For him, all life is descended from a single primordial life in ancient times. DNA evidence supports this idea. But he introduced a believable mechanism called natural selection (adaptation) through which a species preserves an advantageous characteristic that enables it survive while those that are inferior (lack a functional advantage) are gradually eliminated over time. Desirable traits (adaptive capacities!) are passed to their offspring and then inherited by subsequent generations, becoming dominant among the population through time.

According to Freud[4], life is a small detour to death. Life is periodically destroyed by naturally occurring events. On earth, there have been numerous major extinction events that destroyed the majority of complex species alive at the time. The extinction of the dinosaurs is the best known example. Extinctions are believed to be caused by events such as impact from a large meteorite, massive volcanic eruptions, or astronomical events such as gamma ray burst.

---

[4] Sigmund Freud (1856-1939) originated psychoanalysis. For him, people's lived experience is simply the outcome of the conflicting desire to live and the fear of death. People are principally influenced by childhood experiences, cultural environment and irrational drives deep in their unconscious.

Life is a non-spatial property of nature. All life forms come from non-life. Life is thus associated with the ordering principle of the cosmos itself. The universe is alive akin a super-organism. Life is the attribute that distinguishes living from inanimate objects. Though life exists on earth and nowhere else, the universe is an organic whole, a whole in which all parts are intertwined. Life on earth is just but a small part of totality. Biotic factors (all plants, animals and micro-organisms) are continually engaged in a highly interrelated set of relationships with every other element constituting abiotic factors (non-living physical components) of the environment. Indeed, many processes in the Earth's surface essential for the conditions of life depend on the interaction of living forms, especially microorganisms, with inorganic elements. These processes help regulate temperature, atmospheric composition and ocean salinity.

Death

Science revolves around the human desire to preserve life and avoid death. The minimalistic definition of death is the permanent cessation of all biological functions that make life possible or life processes that sustain a living organism or cell. Death would seem to refer to either the moment at which life ends, or when the state that follows life begins. After death, the remains of an organism become part of the biogeochemical cycle. Fossils are the preserved remains or traces of animals, plants, and other organisms from the remote past. Life can be traced to fossils more than 3.4 billion years old.

Contemporary evolutionary theory sees death as an important part of the process of natural selection. Death occurs because during old age our cells begin to slowly regenerate before stopping completely. Orthodox evolutionary models of aging predict that all species eventually age. Aging increases an organism's vulnerability and ultimately leads to its death. Some observation, however, suggest that several species of fishes, amphibians, and reptiles fail to show signs of aging (reduced survivorship and reproductive output with age). Nonetheless, all organisms eventually die of diseases, accidents, predation, etc. It is considered that organisms less adapted to their environment are more likely to die having produced fewer offspring, thereby reducing their contribution to the gene pool. Their genes are thus eventually bred out of a population, leading at worst to extinction and, more positively, making the process possible, referred to as speciation. Extinction is the gradual process by which a group of taxa or species dies out, reducing biodiversity. The moment of extinction is generally considered to be the death of the last individual of that species. Over the history of the Earth, over 99% of all the species that have ever lived have gone extinct.

The nature of death has for millennia been a central concern of the world's religious traditions, philosophical and scientific inquiry. Many religions maintain faith in either some kind of afterlife or reincarnation for the soul, or resurrection of the body at a later date. Mummification or embalming is also prevalent in some cultures to slow down the rate of decay. In most parts of the world dying is dreaded because it is uni-dimensional and unidirectional. Time marks the interval between life and death or the state of being which begins with conception, birth or germination, and ends with death. Even in academic arena, various scenarios of human extinction have been discussed in science. Human extinction is the end of the human species. The scenarios include global warfare annihilation (i.e. nuclear/biological/conventional), overpopulation, global pandemic involving an antibiotic-resistant bacterium, antifungal-resistant fungus, or antiviral-resistant virus, large scale volcanism or other catastrophic climate change, global suicide attack, environmental collapse, loss of a breathable atmosphere (e.g. due to an anoxic event), extreme ice age leading to prolonged global drought, loss of natural resources, such as mass deforestation or contamination of all fresh water, tipping points in climate systems, evolution of another species that out-competes humans for food, habitat or hunts as prey, a geomagnetic reversal, voluntary extinction, scientific accidents or scenarios of extraterrestrial origin.

Life and death are the same. Opposites cohere. The physical world was, is and will be. A thing is dead, dying or alive. All said the physical world is simultaneously present. Life and non-life are connected in the present. Life on earth in the form of humans, animals, plants and microorganisms (the earth's biosphere) is connected to the earth system through biogeochemical cycles. All living things return to dust and ashes when they die, or, to put it another way, to constituent atoms and molecules of hydrogen, oxygen, carbon, phosphorus and so on. But, in another sense, living things do not die: they begin again, from a tiny cell, and scavenge the dust, the air and water, to find the elements necessary to fashion another life. All life is infinitely chained.

Death can and will be overcome. Although death is an integral property of the universe it can be overcome. The cure for death is part and parcel of the universe. Almost all medicines are derived from natural sources. The use of plants, animal products and active chemical drugs (like arsenic, copper sulfate, iron, mercury, and sulfur) as medicines to treat disease and common disorders and as dietary supplements predates written human history and is almost universal. Living life has tapped on the medicinal properties of plants, fungal and bee products, minerals, shells and certain animal parts. The use developed in part as a response to the threat of air, water and food-borne pathogens and other predators such as insects, fungi and mammals. Living life has managed to extract chemical compounds from their surrounding that are used to perform important biological functions, have beneficial effects on long-term health when consumed or administered, which can be used to effectively treat diseases, and/or prolong life. Nonetheless, life forms strongly believe that death is not an unavoidable non-spatial property of the physical world.

Time

Time marks the interval between life and death, and the state that follows the end of life. Time has been a major subject of religion, philosophy, and science. Parmenides maintained that time, motion, and changes were illusions. Noticeably, every event has the characteristic of being both present and not present (i.e. future or past). Gottfried Leibniz and Immanuel Kant held that time is neither an event nor a thing (object/substance); instead, time is part of a fundamental intellectual structure (together with space and number) within which humans sequence and compare events. According to Martin Heidegger we do not exist inside time, "we are time". For him, subjective experience and activity (i.e. mind/time) cannot be made sense of in terms of Cartesian (or material) substances.

Einstein resisted imagining the scenario where the universe is detached from physical content (matter, objects). According to Einstein's conceptualization, space, time and gravity have no separate existence from matter. For Einstein, time and space are a single entity. But time is temporal and unidirectional. Accordingly, with gravity and solid obstacles permitting, it should be possible to move freely through space in any direction. One can go backwards or forwards in the three spatial dimensions. But time doesn't share this multi-directional freedom. The temporal position of events with respect to the transitory present is continually changing; future events become present then revolve into the past. Time is a one way trip in a single predetermined direction: from the past, through the present, toward (into) the future. This is what is commonly referred to as the arrow of time. Traveling to the past is impractical though theoretically possible.

Historically, the concept of entropy evolved in order to explain why some processes occur spontaneously while their time reversals do not. Systems tend to progress in the direction of increasing entropy. Entropy only increases. Entropy never decreases even in seemingly isolated systems. Time moves forward because of the level of disorder (entropy). The arrow of entropy has the same direction as the arrow of time. Seemingly time is irreversible. Nonetheless, thinking beings have long speculated on the possibility and desirability of timeless existence (eternal life).

Things change, seasons change, and people change – simply put it appears that the physical world change. Time is an intrinsic feature of the physical world. Time helps humans to grasp the physical world. Time is a way of separating physical events from each other. Even then, the physical world is a seamless whole.

Time is a vital characteristic of the physical world. Time imitates the cyclicity of the seasons. It is based on the earth's rotation around its axis around the sun as referenced in the form of seconds, minutes, hours, days, months, years, decades, millennia and centuries. Earth's rotation on its axis is directionless. Therefore, time has no direction (contra to Einstein's unidirectional time; that the flow of time is always into the future). It is humans who assign direction to time: circular, wave, spiral, sinusoidal and linear. According to J. M. E. McTaggart's *The Unreality of Time* (1908), time is unreal but events can referred to in terms of 'tensed time': yesterday [past], today [present], and tomorrow [future] or 'untensed time': Monday, Tuesday, and Wednesday. For those who believe in the present alone tensed time is elemental while untensed time alone is not satisfactory. Buddhists, for example, suggest that only the present is real. They consider everything past, future, imagined or mental, unreal.

Linear time denoted by past, present and future is the most common manner for ascribing time to describe the physical world. But why do we remember the past and not the future? John Mbiti (1931-) in *African Religion and Philosophy* (1969) suggests that the future is not knowable. Future time is not knowable because it is not marked by actual events. Time describes physical events that happened, are happening or are just about to happen. The future is unpredictable, because it hasn't happened yet. This is a fallacious thinking because religion by definition imagines a past, present and future (read immortality). Religion is futuristic.

On basis of our understanding of the past and present universe, we can speculate about its future. That is why most religions have a linear view of time characterized by a beginning and an end. Thus life moves from the present to the past. Many religions predict the end of the universe while still demonstrating that human beings never willingly accept their fate. Christianity predicts the end of life on Earth, while human life continues in a paradise. For Christians heaven is the final abode of the virtuous. Hinduism, Buddhism and Jainism, believe in an unending cycle of apocalyptic destruction and re-creation of the universe. In Islamic faith the most valued level of heaven is paradise or Eden (*Firdaus*) to which the prophets, martyrs and other pious people will go at the time of their death.

The linear view of our past in mainstream science presents the picture of the past in which humanity started from primitive beginnings and then steadily progressed with the development of culture and science. However, new evidence present a prehistory characterized by the existence of advanced civilizations before any of the known ancient civilizations came into being. Attempts to explain ancient enigmas such as mysterious artifacts, ancient airplanes, sphinxes of Egypt, the stone building of Axum in Ethiopia and the ruins of Carthage and Zimbabwe, the rock paintings of Egypt, Aztec and Mayan calendar and Celtic cross among others seem to confirm ancient legends and stories describing lost civilizations that perished in a global cataclysm.

Plato in his dialogues *Timaeus* and *Critias* penned that once upon a time (9000 years before Plato's time) there had existed a magnificent seafaring civilization which had attempted to take over the world, but had perished when its island sank into the sea – the result of an unbearable cataclysm of earthquakes and floods. This civilization had been called Atlantis, and it had heralded from the Atlantic Ocean. A number of other cultures also have myths about lost civilizations. Nowadays it is frequently agreed that there is a common historical event or real lost civilization at the root of some or all of these legends.

The perfect theory holds that time is more than what clocks measure. The dimension of time is basically one of the properties of nature. Time is an integral property of the physical world. Time is the mental makeup that facilitates comparison and sequencing of physical events. Time is traversable in either direction. Time is neutral – it is not circular, wave, spiral, sinusoidal or linear. The past, the present and the future are merely comparative. Time is one of the twelve (12) non-spatial extensions of the universe.

Thinking Beyond Time

Time travel is a mystery; a puzzle. From early on, it has been suggested that time cannot be past, future and simultaneously present. However, ancient folk tales and myth have incidences of forward time dilation. Backwards time travel is a common feature in contemporary science fiction. Charles Dickens' *A Christmas Carol* (1843) depicts time travel in both directions. For Saint Augustine, God is outside of time and present for all times in perpetuity. Jung felt that both the past and the future exist in the present, in the unconscious. In William Shakespeare's *The Tragedy of Macbeth* (1603), the plot doesn't unfold in a linear fashion rather the atmosphere of evil is present at the start. The witches look forward to the future. The pattern of events (plot) of a story never starts from the very beginning and must always anticipate continuation of events even with the story being brought to closure. The theories of relativity suggest a scientific basis for the possibility of backwards and forwards time travel in certain unusual scenarios: the use of cosmic strings and black holes, wormholes and alcubierre drive. For Stephen Hawking, time would end with black holes. Square Kilometre Array (SKA) project involves the building of the world's biggest and most advanced radio telescope to be hosted in South Africa, New Zealand and Australia Stars that will enable humans to view the universe as it was moments after the origin of the universe approximately 13.75 billion years ago.

Time dilation according to Einstein is negligible for common speeds, but it increases dramatically when one gets close to the speed of light. Distances appear to contract while clocks tick more slowly when moving at velocities close to the speed of light. Accordingly, time passes more slowly the closer you approach the speed of light, an unbreakable cosmic speed limit (when you reach the speed of light time stands still). If you travel faster than the speed of light it is assumed you can go back in time. The twin paradox is a hypothetical case where a twin makes a journey into space in a high-speed rocket and returns home to find he has aged less than his identical twin that stayed on Earth. Einstein invoked gravitational time dilation to explain the aging as a direct effect of acceleration. But speed isn't the only factor that affects time. Mass also influences time. Time is dilated by matter. Time slows down the closer you are to the center of a massive object. Time runs a little bit faster in zero-gravity space than it does down on Earth. According to Einstein's theory of gravitation, any object with mass will cause a warp in space-time, similar to a bowling ball on a mattress. Because space and time has been stretched, clocks operate slower close to Earth than in the vast areas between galaxies. Therefore, relativity predicts that if one were to move away from the Earth at relativistic velocities and return, more time would have passed on Earth than for the traveler, so in this sense it is accepted that relativity allows "travel into the future".

The entire universe, as we understand it, is beholden to the rule that something occurs first and the outcome of that occurrence happens afterward. So, the speed of light is also the cornerstone of the concept of causality: causes precede effects, anywhere. The insinuation is that it is highly unlikely to travel faster than the speed of light. But if one could travel faster than the speed of light then time would, theoretically reverse direction and the person could arrive before s/he departed. Cause-and-effect would be undermined. The grandfather paradox is an imaginary situation in which a time traveler goes back in time and attempts to kill her grandmother at a time before her grandmother met her grandfather. If she did so, then her mother or father never would have been born, and neither would the time traveler herself, in which case the time traveler never would have gone back in time to kill her grandmother. There is also the case where a time traveler goes back and attempts to kill herself as an infant to pre-empt her own existence. If she were to do so, she never would have grown up to go back in time to kill herself as an infant. Such a possibility raises provocative questions about the meaning and randomness of our existence.

There are two major objections to this line of thinking. First, the multiverse model suggests that there is not one universe but many. The implication is that every possible outcome of any event actually takes place. In this multiple universe, a time traveler who goes back in time to murder their own granny can get away with it because in the universe next door the granny lives to have the parent to the time traveler who becomes the murderer of their grandparent. Secondly, David Hume in *A Treatise of Human Nature* (1739) claimed that causality could not be observed empirically. He suggested that causality is based on experience, and experience similarly based on the assumption that the future models the past, which in turn can only be based on experience – leading to circular logic. He asserted that only correlation can actually be perceived. For him, correlation does not imply causation.

To date time travel, as in moving between different points in time as fast as or faster than the speed of light, is the last frontier of science. Indeed when it comes to time travel, the lines between science fiction, fantasy and horror are thin. Even then, by the time H.G. Wells' *Time Machine* (1895) was written, imagining the invention of the car, airplane and other modes of transportation seemed nothing short of mental illness. It is on the basis of this that some scientists such as Stephen Hawking strongly believe that time travel into the future may become a reality someday in the future. Ironically, Stephen Hawking once suggested that the absence of tourists from the future is enough evidence in an argument against the existence of time travel. Later, he admitted that he had avoided talking about time travel previously 'for fear of being labeled a crank', saying the subject had once been 'scientific heresy'. Hawking offers the view that humans will be able to travel millions of years ahead of their own time by means of a time machine.

According to *The Perfect Theory*, time travel is not physically possible. However, simultaneity (the capacity for someone to travel from one part of the universe to another in an instant) is probable except that space (the physical world) is not empty. The physical world is always simultaneously present. Though Einstein's theory of relativity states that light is the fastest matter in the universe, instantaneous (faster than light) communication between events or particles in the physical world is conceivable. Ponder this. If someone is killed in one location and a close friend who is in a far-flung location senses that something is wrong with them, the question arises, "How does information get around so quick?" This is a case of connectedness. The event of death (the fracture of the special bond between two objects sharing the property of life) is communicated to the friend instantly as a loss.

Results of tests of Bell's theorem (named after John Stewart Bell (1928-1990) in *Speakable and Unspeakable in Quantum Mechanics* (1987) appear to demonstrate that some quantum effects travel faster than light. This is a pointer to the reality that apparently separate particles (and events) can influence each other instantaneously over great distances. For instance, polarized light are correlated, no matter how far apart the particles are. On the same note, the Backster Effect showed that animals and plants communicate and that communication is immediate. Backster discovered that plants display emotion in life-threatening situations – plants are able to read the human mind before the implementation of the threatened act and react accordingly. Judgment of time can also be also altered by physical alterations to the brain such as with traumatic brain injury, psychoactive drugs, temporal illusions, age, hypnosis and neurological diseases such as Parkinson's disease and attention deficit disorder. For example, under hypnosis, some patients allegedly, have vivid memories of past lives. During the ordeal, they are capable to travel to other locations.

The enterprise of science is based on the notion that the universe is governed by the impersonal principles. For instance, events in history move in obedience to certain general laws – man is destined to die. But applying the principle of unity of opposites, this perception has an opposite rule stating that the principles governing the universe can be tinkered with. The preoccupation of scientists is to grasp fundamentals of nature so that they can tinker with them. Scientists have been pondering on the likelihood of travelling at the speed of light. Such an achievement will mean remaining in effect remaining ageless because then time stands still. It is possible human beings aspire to experience the past, the present and future all at once. Eternal life is timeless existence.

The gist of *The Perfect Theory* is to grasp how the content of the human mind shape our experiences and also the nature of experiences and how the mind deals with such experiences. Human beings desire to actualize the idea of travel in time. Immortality summarizes all meaningful thought about the world. Humans imagine a situation where they will never get old instead they would stay young forever. They can talk to the future and to the past. This quest is not just limited to rational enquiry but rather the interaction of all living life with the physical world. Therefore, consciously or unconsciously all living life is involved in an attempt to undermine the inescapability of their fate.

Arthur Schopenhauer (1788 – 1860), a contemporary of Hegel, in his most influential work, *The World as Will and Representation* (1818), divided the world into will (life instinct) and representation. He believed that we could gain knowledge about the thing-in-itself. Will, for Schopenhauer, is what Kant called the "thing-in-itself." For him, self-preservation is the overriding motivation among human beings and all observable phenomena. Therefore, the entire universe and everything in it as driven by a primordial desire to avoid death and to procreate. The pessimist Schopenhauer concluded that human desire and action is illogical, directionless and self-defeating. He saw in art and aesthetics especially music and aesthetic living a temporary way to escape the unavoidable. Yet probably no one escapes the ravages of time.

Ancient Greek thinkers believed that the universe is infinitely old, has no beginning and extends infinitely into the future. The theory of temporal finitism inspired by the doctrine of

Creation and advocated by medieval philosophers and theologians developed the concept of the universe having a finite past with a beginning. This view is shared by Abrahamic faiths (Judaism, Christianity and Islam) as they believe time started by creation. Time is finite, that is, time will end with the end of the world. Therefore, at the center of religion is the attempt to grasp the physical world and the possibility of demise. Most religions include a belief that at some time in the distant past, man dwelt happily with God, but that a separation took place resulting in viscidities of life (especially death). The Big Bang theory states that the universe is limited in extent (space) and time. The universe began in a colossal violent implosion. The current estimate for the age of the universe is 14 billion years. Many possible ultimate fates of the universe are predicted by rival scientific theories, including futures of both finite and infinite duration.

Immortality is the ability to live forever, regardless of whether or not the body dies. Belief in a future life of some sort seems to have been practically universal at all times. However, it is unknown whether human physical immortality is an achievable condition. Natural selection has developed potential biological immortality in at least one species, the jellyfish (*Turritopsis nutricula*). The jellyfish has an indefinite lifespan. Most other biological forms have inherent limitations which may or may not be able to be overcome through medical interventions or engineering. For example, there is apparent absence of aging in bacteria which reproduce through cell division. However, bacteria as a colony may eventually die since each succeeding generation is slightly smaller, weaker, and more likely to die than the previous. The Bristlecone Pines (*Pinus longaeva*) have the capacity to recurrently create biological or synthetic replacement parts to replace damaged or dying ones.

Stephen Cave in *Immortality: The Quest to Live Forever and How It Drives Civilization* (2012) observes that the search for everlasting life has been one of humankind's major preoccupations since the dawn of humanity. He suggests that this quest to overcome mortality is the driving force behind human civilization whether human beings are conscious of it or not. The pursuit of immortality takes four forms: (1) staying alive – prolonging physical life through medical or technological means; (2) resurrection of the body as in return to earth after death but in another form as is the case with all Abrahamic/monotheistic religions (Judaism, Christianity and Islam); (3) Reincarnation or the continuity of the immaterial soul based on the idea that our individual consciousness survives physical death as suggested as early as in Plato's time; and/or (4) legacy/popularity as in becoming famous through grand deeds so that one's name continues to live on, transcending time and distance. On the whole, Cave concludes that the appeal of living for ever (even imagining it is probable) is mere fantasy. There is no life but this one.

Certain scientists, secular futurists, and philosophers intimate that bodily immortality is achievable through life-extending substances, human cloning, cryonics. Others believe that life extension (absence or delay of human aging) is a more achievable goal in the short term. Ray Kurzweil and Terry Grossman in *Fantastic Voyage: Live Long Enough to Live Forever* (2005), advocates for life extension in anticipation of the possibility of humanity overcoming all diseases, ageing and senility. However, the reversal of aging would not guarantee humans biological immortality for they can still die from physical trauma. Perhaps the hope that one day humanity will be able to live forever seems farfetched. Even then, the form an unending human life would take, if it were to become a reality, remains the subject of speculation, fantasy, and debate.

Why Do We Mourn the Dead?

Mourning is a complex range of behaviours to express unhappiness over the death of someone. It is the public display of unrestrained emotion of sorrow in the event of death. Apparently mourning is a universal impulse. Diverse reasons are offered for the ritual of mourning including an opportunity to let go of the memory of the departed, emotional attachment to the deceased, respect for the dead, absence makes the heart grow fonder, good memories of shared times which will not happen again and loss of loved ones particularly family members, friends and or neighbors. But how come people mourn total strangers or even enemies?

The mourning behavior and burial rituals performed by Ancient Egyptians is illustrative. Divine kingship was the basic political institution of ancient Egyptian civilization. The Egyptians believed their king (or pharaoh) to be both a god and a man. He was regarded as a benevolent protector who controlled the flood waters of the Nile, kept the irrigation system in working order, maintained justice in the land, and expressed the will of the gods by what he said. It was expected that when their all-powerful god-king died and joined his fellow deities, he would still help his living subjects. Pharaohs were regarded as having control over the journey of life. Egyptians had effective methods of embalming bodies of their pharaohs to keep them intact for thousands of years. George Allen in *The Book of the Dead or Going Forth by Day: Ideas of the Ancient Egyptians Concerning the Hereafter as Expressed in Their Own Terms* (1974) observes that Ancient Egyptians maintained an elaborate set of burial customs that they believed were necessary to ensure immortality after death. These customs involved preserving the body by mummification (of King's skin, bones and internal organs), performing burial ceremonies, and interring with the body goods the deceased would use in the afterlife. Wealthy Egyptians were buried with larger quantities of luxury items, but all burials, regardless of social status, included goods for the deceased.

Shaka, the legendary Zulu king of South Africa, was so overcome with remorse that he ordered his subjects to mourn for two years after his mother, Nandi, died of dysentery. He put on his war regalia and proceeded to wail and shriek in a public display of unrestrained anguish. Tradition held that upon the death of someone of Nandi's stature, a range of personal attendant would be wounded or killed. On Shaka's orders, several people were put to death on the spot, and a general mass execution broke out. More than 7,000 people died in the massacre. Shaka himself was killed later the same year by his half-brother Dingane. As custom demands his body was buried with all his possessions. Several servants and attendants were killed and buried with the deceased as per tradition people of their rank could not die alone for they required servants to attend to their needs.

Many mortals have been immortalized by having public institutions, roads, streets, places etc. named after them. After many countries in Africa attained self-rule, Kwame Nkurumah, the champion of a united Africa (Pan-Africanism) in order to break the shackles of neo-colonialism wished his body would be embalmed and preserved like Lenin (Russia's communism icon).

Historically, some named groups of people did not always make elaborate plans for the dead and the dying. Some did not care what happened to their dead and routinely threw them in the bush where hyenas and other scavengers could devour them. Accordingly, the bushes served as the open communal graveyard. Ordinary people who died or were terminally ill were merely thrown into the bush to be eaten by wild animals. For some it was considered bad luck for a person to die inside the house and when this happened the homestead was cleansed and the shelter would later be burnt down. Today most people bury the dead hygienically in solemn ceremonies with or without fanfare. Many people still inter the deceased together with their clothing, bedding, furniture and food items to make sure the deceased's after life was comfortable. Many people also conduct cleansing ceremonies to cleanse those who had dug the grave and handled the body before they could be fully allowed to socialize with other members of society.

So what exactly are we to be afraid of? - Death. All over the world, death is dreaded. According to Mbiti, in some parts of the world, the dying process is referred to as crossing over or passing over instead of just the end of life. Death is experienced as a group event. Family and community experience a transition as a member 'passes over' to the other world. Death is seen as a rite of passage wherein the soul passes into another phase of continuous existence; the soul leaves the material world and crosses over into the spiritual world. Death is part and parcel of the rhythm of nature that consists of birth, puberty, initiation, marriage, procreation, old age, death, entry into the community of the departed and finally entry into the community of spirits. Medicine men represent hope, since they have the power to reverse misfortune by curing disease. The same can be said of people with special power such as rainmakers, mediums, medicine men, witch doctors, and tribal chiefs.

Human beings aspire for power, material possessions, comfort, popularity, and the pleasures of the senses. They are deeply engrossed in the pursuit of knowledge, beauty, love, power and money. Humans live their lives as if they are going to be in this world forever. Probably human beings in their hearts of hearts never desire to die. Mourning is the most conspicuous expression of the human mission to safeguard life and circumvent death.

Why Do We Study Science?

For ages scientists have passionately argued that the laws that govern physical phenomena are completely different from those that apply to human activities and social environments. However, this position appears to falter when you consider two fundamental questions in science: (i) what is science? and (ii) why do we study science?

As a field of study science refers to a system of acquiring knowledge to describe and explain the universe. Science is a systematic enterprise that builds and organizes knowledge in the form of testable explanations and predictions about the universe. The term science also refers to the organized body of knowledge people have gained about the physical world and its phenomena using the scientific method (unbiased systematic observations and experimentation). Therefore, the word science often describes any systematic field of study or the knowledge gained from it. Fields of science arising thereof are: Natural sciences (study the natural world) and social sciences (the systematic study of human behavior and society). Central to the division is the contention that the kind of knowledge that is acquired by the observation and investigation of the nonhuman or natural world is of a radically different type from the knowledge that humans can achieve of their own actions, creations, and institutions.

The purpose of science it is said is to produce useful models of reality or the operation of fundamental laws. Nevertheless, this perspective may not be entirely correct. Science is not merely an attempt to grasp the physical world. Historically, alternatives to mainstream science have often been sidelined, ignored and dismissed as non-scientific. The object of science includes the practical application of research (scientific findings) to satisfy human needs and desires and to address their fears (applied science). Therefore, science is knowledge acquired through study and practice; science is both an academic endavour and a practice. This view of science is a synthesis of oral traditions (theological tales and myths, legends, the occult manipulations of alchemists, the astrology, magic/sorcery/witchcraft, miracles, mystics, supernatural powers (gods/God), child play, music, art, poetry, literature, activities of politicians and ordinary people and the advancements made by modern day scientists.

Science has withstood numerous challenges and difficulties in the tortuous journey to be what it is today. Science started with religion and that is why religion was a crucial factor in the rise of ancient civilizations. The world's first civilization arose some five thousand years ago in the river valley of Mesopotamia (later Iraq). Religion and myth were the central forces in these early civilizations. They pervaded all phases of life, providing people with satisfying explanations for the operations of nature and the mystery of death and justifying traditional rules of morality. The notion of supernatural phenomenon and deities marks a decisive transition in human awareness that they are not condemned to be mortals forever. So, religious people were the first humans to imagine infinity. This train of thought appears in oral traditions (myths, epics, legends, riddles, proverbs, fables, allegory and stories), witchcraft, miracles, early alchemists and of course modern science as an academic endavour and practice.

Oral traditions were limited in actualizing time without end because they were confined in spread to membership in particular cultural traditions and often they were passed on from one generation to another without much improvement. So, with time they became mere customs and traditions. For example, Evans-Pritchard (1902 – 1973) in *Kinship and Marriage Among the Nuer* (1951) reports that among the Nuer, every person is viewed as part of a larger whole. The relationships between the human and the non-human, the visible and invisible, and the living and the dead allows for a unity to exist between the dualistic realities that are often polarized. The dead ancestors are considered to be very much a part of the living. Dead persons, though for the most part only men, are remembered in their names. A person gains immortality through the remembrance of their name. Every man was expected to have at least one son who will carry on his name and through whom his name is forever a link in a line of descent. The Sumerian *Epic of Gilgamesh*, the oldest known literary work on earth, dating back at least to the 22nd century BC, is primarily the quest of an ancient adventurer king to become immortal. The quest to live forever also features in the Homeric poems and the writings of the Old Testament where Adam and Eve were thrown out of Eden for fear that they may eat of the tree of life in what is commonly referred to as the Fall of Man. Paramahansa Yogananda (1893 – 1952) in *Autobiography of a Yogi* (1946) narrates stories of immortal saints such as Mahavatar Babaji.

Witchcraft is secretive and is often considered anti-society. For centuries, human body parts have been used as ingredients for magical concoctions and charms. To obtain body parts, performers kill people in order to harvest specific organs for use in the occult. Therefore, knowledge of witchcraft cannot be passed on smoothly or improved on. Practitioners sometimes resort to trickery and short-term ends further bringing the profession into disrepute. Miracles are also limited because their performance is confined to a select group of individuals with special powers. Each and every individual is religious. This is because the religious instinct is also the preservation instinct. No one ever wants to die.

Individuals are members of cultural traditions that often project the preservation instinct in the name of the group (cultural exclusivism). Group conflicts arise because their interests appear to conflict. For example, the Old Testament is an account of the conflicts between the Hebrews versus non-Hebrews. Christianity emerged during the first century of the Roman Empire. Jesus' life, teachings, crucifixion, and the belief that he had risen from the dead convinced his followers that Jesus had shown humanity the way to salvation. Dedicated disciples spread this message throughout the Mediterranean world. They soon spread out as missionaries to Jewish and Gentile communities throughout the Roman Empire and beyond.

Today religion is used to settle scores among mortals limiting its chances of taking humanity to a higher level. Religion is a contestant in ideological, civilizational and religious conflicts. Even global terrorism is motivated by religious jingoism.

Early alchemists gave birth to physiology and later modern day medicine. Academic scientists explicitly reject divine explanations for human occurrence, search for natural cause, and base their conclusions on evidence. The advantage academic science has is that most humans are free to partake in it.

Generally academic science, spiritual traditions and activities of everyday life together offer a remarkably compatible view of the universe. Science in a holistic sense is the acquisition, development and deployment of human adaptive capacities to preserve life and avoid death. Science is motivated by the struggle to manage life's greatest misfortune – death. The predicament of science is death. The urge to preserve life is the reason as to why living life walk, climb, procreate, drink, eat, sleep, kill, commit suicide, pray, meditate, think, invent, etc.

CHAPTER THREE

HOW THE TWIN IDEAS OF LIFE AND DEATH STRUCTURE SOCIETY

Human Adaptive Capacities

Consciousness is the state of being aware of one's surroundings. Charles Darwin (1809 - 1882) noted that all organisms constantly adapt to their environments to endure. Evolution operates through natural and sexual selection mechanisms. All human beings are endowed with morphological adaptations such as hearts, mental organs, differences between sexes, and physical characteristics. Nevertheless, it is almost certainly false to characterize life in terms of physical processes alone. Human beings are also endowed with psychosomatic adaptations particularly mental faculties, senses and emotions.

Mental properties incorporate reason, imagination, perception, intuition, memory and will that in turn shape beliefs and desires. Mental faculties are necessary components in the evolution and survival of the human species as they enable humans to solve complex problems simply by thinking about them and envisioning the possible outcomes of specific actions.

The human capacity for the imaginative construct is apparent in human social-political institutions (family, lineage and descent groups, clan, ethnic group, the state, religion, race and civilization) and culture (symbols, myths rituals, clothes, use of tools etc.). Emotions can be observed in empathy, compassion, mourning, cooperative aspects, altruism, aggression and nurturance.

Consciousness is the product of sense experience and mental activity. For years, the human mind has had tremendous effect on the surrounding material world in the quest to sustain life.

Knowledge is the world's most valuable resource. Knowledge facilitates adaptation. Human socio-political institutions and culture are adaptive capacities. That being the case, struggles in the human society are not attempts to gain access to and control of limited resources - the social (e.g., mates), biological (e.g., food), and physical (e.g., territory) rather humans are engaged in the pursuit to perpetuate life and eschew bereavement.

Social behaviors such as mating patterns, territorial fights, pack hunting, slavery, colonialism, imperialism and war are probably efforts to safeguard life (and avoid death). Life (bonds of hope) and death (bonds of fear) structure the human society.

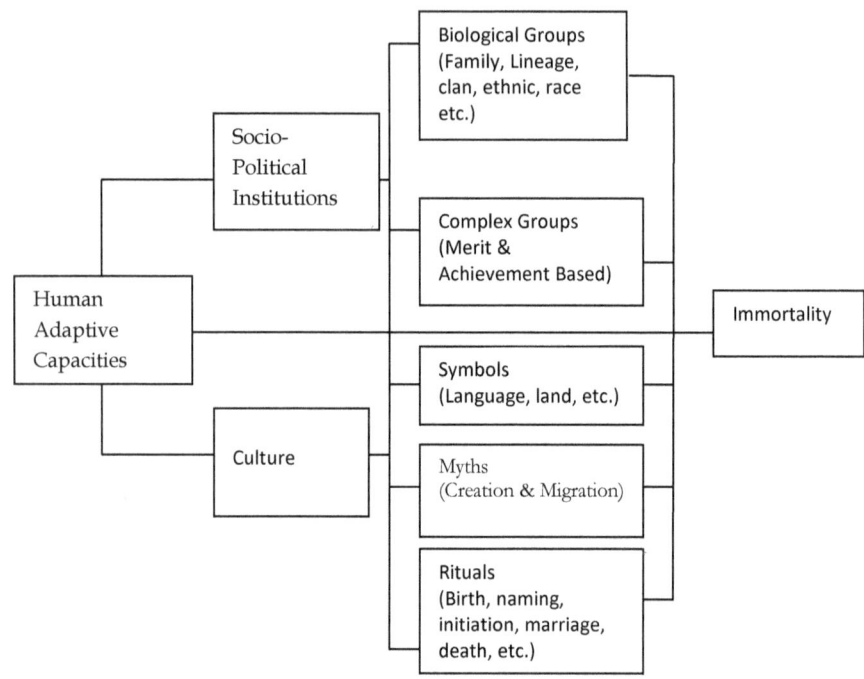

Figure 4: Human Adaptive Capacities

Underlying *The Perfect Theory* is the understanding that behaviors have evolved over time, similar to the way that physical traits are thought to have evolved. The theory predicts, therefore, that human beings will act in ways that have proven to be evolutionarily successful over time, which can among other things result in the formation of complex social-political institutions and processes conducive to the perpetuation of life. However, in the contemporary world well-intentioned behaviors seems to have gone awry. Why?

The quest to sustain life and eschew death structures the human society. Markers of identity constitute critical dimensions of social organization. The important markers of identity include the family, kinship/lineage, clan, ethnicity, state, race, civilization, religion, status, caste, age and principally gender.

Gender structured units (hierarchies in the human society) are a means of adaptation to externalize conflict and solidify human groups ranging from the family, lineages, clans and ethnic groups to face off external aggression. The taboo on incest, which is almost universal, helps to avoid competition and conflict within the family.

The survival of the family rests on the restriction of sex among its members and the consequent need for other family units as sources of spouses. The taboo normally extends to close (and sometimes distant) relatives for the same purpose thus bringing into its fold lineages and clans (especially the clans that ascribe to exogamy). Thus faced with possible external threats and competition for mating partners, structures such as the family, lineages, clans, tribes, ethnic groups, etc. are constructed in such a way that they ensure the survival of humans. However, gender informed identities only attempt to project competition and conflict on the outside but fails to eradicate such competition and conflict entirely. Therefore, despite the externalization of enmity, people compete, conflict and die within and between the various constructed socio-political identities with either gender (mostly men) projecting the strength of the group through competition, conflicts, slavery, colonialism, imperialism, concentration camps, terrorism, war and genocide.

Adaptation in Our Time

From the preceding thought process, we know that human beings have non-spatial extensions notably mental faculties among eleven (11) others. Mental qualities incorporate intellect, reason and understanding. Works of art, music, architecture, and literature are products of our mental faculties. Many traits of human mental faculties are also apparent in human socio-political institutions, linguistic capabilities, programming (coding) and scientific breakthroughs.

Human societies are characterized by patterns of relationships between individuals (and institutions) that enable individual members to achieve personal needs or wishes that they could not fulfill separately by themselves. Human social organization is an adaptive capacity arising from the way human beings interact with their environment. In the long-term, humans wish to establish a perfect society – a fair society. A fair society is only possible if and when self, selfless and collective interests are reconciled. Social groups emerge because of the perceived differences between the personal and the cooperative. Such competing interests (discrimination and persecution of social groups) can make it difficult in the fulfillment of broader societal goals. Politics (and political reforms) provides the alternative opportunity for bringing together these manifold interests.

Politics largely deals with participation and representation in decision-making in the process of the allocation of value. Politics has a domestic and international facet. International politics is framed around transnational state actors and non-state organizations that may include Non-Governmental

Organizations (NGOs), Multinational Corporations (MNCs), terrorists, etc. Political science is the methodical study of the theory and practice of politics. Political science methodically analyses political systems and political behavior. Political scientists dispassionately originate the analytical frameworks from which politicians, the electorate, journalists and advocacy groups scrutinize political phenomena.

For thousands of years thinking gurus have pondered over the principle and practice concerning community, public life, and social organization.

Socrates (469-399 B.C.) suggested that the dialogue was the sole avenue to moral insights and self-knowledge. Socratic thought held promise of reforming the individual through the critical use of reason. In Plato's Euthyphro (399 BC), Socrates called the old gods and the old laws into question and asked others to challenge the fundamental beliefs upon which their society rested. When Socrates was seventy, he was accused of corrupting the youth of Athens and of not believing in the city's gods but in other new divinities and he went on trial for his life.

But Plato (423-347 BC) felt that the individual could not undergo a moral transformation while living in a wicked and corrupt society. If the individual were to achieve unadulterated virtue, the state had to be reformed. In *The Republic* (380 BC), Plato proposed organizing government in harmony with the needs of human nature and informed by the principles of reason. According to Plato we live in an orderly universe informed by natural law (contrasted by positive or man-made law). Natural law implies rules of conduct universally applicable based on reason.

Aristotle (384 BC – 322 BC) in *Politics* (350 BC) suggested that humankind's natural habitation was the state (polis). He advocated for a mixed constitution (government). For Aristotle, a man identifies himself first and foremost as a citizen of a particular polis or city. He would then be counted on in defending the polis from attacks, sustain its institutions of justice, and contribute to its common good. In this way, his own pursuit of a good life is inextricably bound to the fate of the polis. Slaves, women and non-Athenians (or 'barbarians') lacked any entitlements. Yet the Greek term cosmopolis represents unsolved tension between the cosmos (a natural universal order) vis-a-viz the polis (one's local city-state).

St Augustine (354 – 430 A.D.) in the *City of God* distinguished between the city of man and the city of God. He argued that one was not a member of his city but rather the city of God. The city of man was a sinful place. He suggested that God was the source of all knowledge. As such reason by itself was an inadequate guide to knowledge without the revealed truth. St Augustine's was largely a rewrite of Plato in a Christian context.

One of the most influential works during the early modern period (renaissance) was Niccolò Machiavelli's *The Prince* first published in 1532. Machiavelli can be credited as the founder of the modern state having rescued human thought from being too preoccupied with theology and for embracing secular politics. Machiavelli associated politics with power and interests. Machiavelli significantly influenced Thomas

Hobbes (1588-1679), John Locke (632-1704), John Stuart Mill (1806-1873) and particularly Jean-Jacques Rousseau (1712-1778). Ever since Machiavelli, the preoccupation of human beings has been on the establishment of an efficient, responsive and capable state.

Rousseau in *The Social Contract* originated the concept of "general will". For Rousseau, the social contract is the foundation of the will of all people as a whole for the reason that nature itself provides no guidelines for determining who should rule and in whose interests should they rule. Rousseau controversially argued that in a democratic setup the people at large would act as sovereign. Political scientists have ever since been preoccupied with how to organize political systems to sustain the will of the people. The popular will in a world characterized by competing identities regularly fails humankind in the quest to achieve an evenhanded society.

In the quest for an ideal society, humans have experimented with different forms of organizing society including bands, tribes, chiefdoms, kingdoms and state societies (Service, 1975). Presently, bureaucratic state societies have almost completely displaced primordial forms of social organization all over the world. Still, the present state societies are unsatisfactory because they are divided and fragmented along families, lineages, clans, ethnic groups, races, religions and civilizational persuasions. Organized human groups and entities worldwide continue to perpetrate the most heinous crimes out of fear (imagined or real; informed or uninformed). Individuals as members of identities in pursuit of self-preservation have opened room for indiscriminate thoughtless brutality endangering the very life they sought to protect.

105

History is replete with examples of ordinary people who have been turned into murderous lunatics suddenly by thieving despots and a few of their cronies. People are routinely detained, tortured, exiled and even killed for either belonging or not belonging to specific identities. The human insecurity and susceptibility arising from such experiences is such that at present not everyone wishes to be categorized. Perhaps instead of self-preservation the focus should be on collective preservation.

Actually contra to Rousseau, the principle of the universe is to sustain life. Human civilization has not been progressing linearly since the human society is demarcated by boundaries of relatedness[5]. For the most part of human history, people have lived next to people who were like them with each hypothetical biological community having its own neighborhood to itself. Everyone was supposedly related to everyone else. Relatedness is a no win situation. The impulse for self-preservation prompts us to stereotype, discriminate, compete and even eliminate those who we deem to be threats to our survival. However, the natural world is in sync with the will of the universe (or collective preservation) which can only be guaranteed by connectedness. Connectedness is a win-win situation for it mirrors a universal psyche, inclusivity and collective survival.

The allocation of value to seemingly competing interests requires participation. Observably, it is not practically possible for all to participate in decision-making. Democracy is one of the processes by which preference is made on the type of

[5] Politics along fault lines

representation (political structures) and the slate of representatives.

Today, democracy is the most preferred political system the world over in comparison with existing authoritarian alternatives. In a democratic setup it is possible for change of officeholders to occur devoid of killings, force, fraud, pretence and or tribal or family connections. In addition, participation in decision-making is the cornerstone of democratic systems of governance. Democratic political configurations require citizen participation (interest articulation) and competing platforms (political parties) for contenders to articulate and possibly convert their agendas into policy programs (interest aggregation). Baron de Montesquieu succinctly described division of political power among an executive, a legislature, and a judiciary. The three branches of government are largely distinct of each other, each with its own mandate and have checks and balances on each other. In this way, no one branch can acquire unqualified power or misuse the power they are given. Further, democracies enjoy political stability and provide relatively better conditions for doing business. For these reasons, democracy is relatively better at securing and sustaining life in comparison to all others.

Power, the key concept in politics, is a mental attribute. Politics is not a deadly zero-sum game for power and resources. Political power is not strength, control of coercive instruments, manipulation, empire building, colonialism, imperialism or hegemony. Power is not an end in itself. Power is a means to an end. Power is a peculiar form of connectedness. Cooperative power (synergy) is the bridge of life. A meaningful life is the end we all aspire for.

Periodic elections have steadily replaced revolutions, coups and assassinations as the default method of changing governments. Political leadership was never meant to be merely a means of livelihood for mere careerists. Politics is a pedestal for visionaries, those who think beyond the human divisions of their time. Criminals, scoundrels, usurpers and pretenders have often times sneaked their way to power and humanity has paid dearly from their delinquency. Eloquence, charisma and demagoguery only qualify one for leadership in deeply flawed political systems. Sadly, defective political arrangements have dominated the world for the most part of human civilization. It is therefore hardly surprising that the democracy assignment, a process that is now more than 2,500 years old, still remains unfulfilled. Despite steady efforts no country has achieved perfect democracy. Democracy is considered perfect when the theory and practice of politics, besides achieving fairness, inclusivity, representation and participation, venerate the will of the universe (which may not necessarily be the will of the majority). Such a scenario imagines a universalist worldview whereby self and collective interests converge. Social groups and identities demarcate the human society raising the possibility of war (violent encounters over competing interests)

War is the human activity that is associated with killing on command with the hope guaranteeing own survival. In war you kill, get killed or desert duty. Warfare has been a feature of human life throughout history. Archaeologists have found evidence of ancient conflicts everywhere, from Mexico to Mesopotamia. Starting with rudimentary weapons, people have developed ever more advanced methods to scare, maim or kill one another. The surfacing of the state has in some measure mitigated intra-state wars but rendered the occurrence of intra and interstate wars toxic. Identity struggles and the accompanying sectarian divisions, civil strife, conflicts and wars only benefit a cartel of national elites in the short term. In the long-term, war situations are no-win scenarios.

Strategists and promoters of war situations such as Sun Tzu in *Art of War*, Thucydides in *History of the Peloponnesian War*, Chanakya (or Kautilya), Han Feizi, Niccolò Machiavelli in *The Prince*, Thomas Hobbes in *Leviathan*, Frederick the Great, Charles Maurice de Talleyrand-Périgord, Carl von Clausewitz in *On War*, Otto von Bismarck, Henry Kissinger, Charles de Gaulle, and Joseph Stalin were not bothered with the underlying origins, continuation and causes of misgivings, animosities and inequalities between purportedly different groups.

War is not a permanent feature of the modern state system. War is the consequence of suspicion, insecurity or fear of death informed by the urge for preservation (instead of collective preservation). States always arm themselves to secure their domestic and foreign interests because their leaders mistakenly believe government-sponsored defense is the only way to make themselves and their state secure in an environment of anarchy (or in the absence of a universal sovereign or worldwide government. Yet arms acquisition (or deployment) that one state uses to increase its security decreases the security of other states. This results in the situation of the security dilemma (Herz, 1951).

In the real world, there is the self, the selfless and the collective. Connectedness is a combination of the self, the selfless and the collective. War distorts connectedness – violates the edict of the universe. The history of human civilization is proof enough that an idea that is ripe for the dustbin cannot be retrieved. Human beings tend to discard ideas, events and institutions that end up precipitating rather than preventing unnecessary loses of life. Mention can be made of cannibalism, infanticide, killing of twins, human sacrifice, dueling, slavery, sterilization, geronticide and as anticipated war. Currently, we are in throes of the rise of the international system (and the global citizenry). Historical reality and trends indicate that the scope for military influence in world affairs is narrowing (Mueller, 1997). The development of nuclear weapons marked the end of conventional warfare (Sagan &Waltz, 2002). The model state is one which is not armed. This involves the pursuance of a common global defense strategy (collective security) that overrides state boundaries and other edifices of fear. Perfect states do not pursue self-interest because the international state system is not anarchic. The will of the universe (not the will of the people) is sovereign and the foundation of order in the state system. And it is slowly becoming clear that the future of the world is one dominated by networks of inclusive and outward looking neighborhoods that stretch out on a world scale. This way voters guided by a common value system and goals have the power to remove leaders who abuse their trust.

Historians have long speculated whether it is possible to discern the path of human history, and the direction and goal of humankind in the historical process. Some historians suggest that history is directionless (Tolstoy & Popper); others suggest that history is cyclical (Ibn Khaldun, Spengler &

Tonybee); and others opine that history unfolds towards a final destination (Hegel, Marx & Fukuyama). Linear historians disagree on whether the direction of history is influenced by ideas (Hegel), materialism (Marx) or both ideas and economics (Fukuyama). According to *The Perfect Theory*, history makes known the specific destination towards which humanity is going. Historians utilize numerous tools to explore history namely: oral sources, material remains, artifacts, architecture, biofacts and landscapes (archeology), study of language (linguistics), anthropology and written sources. The known past is a reminder of a shared humanity and an affirmation of a future that is worth working toward. Narratives, records and historiographies about the human past help to organize society to fulfill human needs. History is the development of the human awareness of their connectedness. History is the story of humankind in the quest to preserve life indefinitely.

Newtonian time is absolute while according to Einstein time is relative. For Einstein, time runs differently in different planes of existence. But, according to connectedness, the past, present, and future all exist simultaneously. Time is a non-spatial extension of the universe. This explains why scientists scrutinize the most far-flung and oldest galaxies to help them understand the properties of the early universe. Humanity's desire is to be able to tinker with nature (to modify history) through the acquisition of the capacity to move through time into the future and/or into the past.

Immanuel Kant (1724 – 1804) in *Critique of Pure Reason* (1781) famously rejected the traditional arguments in favor of the existence of God. However, in *Critique of Practical Reason*

(1788) he suggested that belief in a supreme deity and immortality is a prerequisite for moral action. In Fyodor Dostoevsky's *The Brothers Karamazov* (1880), the character Ivan Karamazov exclaims: "If there is no God, then everything is permitted… if there is no immortality, there is no virtue". For Plato, we do not subscribe to the idea of immortality for convenience, immortality is real. In Plato's *Phaedo*, Socrates in his final days with his disciples, moments before drinking the hemlock shows no sign of fear or concern, for he is certain that he will survive the death of his body. Socrates appeals to cycles and opposites. He believes that everything has an opposite that is implied by it. And, as in cycles, things not only come from opposites, but also go towards opposites. In the same manner, life and death are opposites in a cycle. We shall have a life after we die. For Archbishop Desmond Tutu, "In some ways, death is like a birth; it is the transition to a new life" (Tutu, 2014).

Human beings are determined to lead a perfect life. Human beings would rather they were able to live an infinity life. Humans mourn the dead out of the knowledge that the same would someday definitely happen to them. They erect monuments in remembrance and defiance of death. It is the human quest to overcome finite life (experience) that gives meaning to human existence. Religion is part and parcel of an intellectual thought process to sustain life (endlessly if possible).

Language is primarily human cognitive capacity that enables humans to learn and use systems of complex communication. Humans communicate and share information through spoken, written and sign languages. Humans also have the capacity for artificially constructed communication systems such as those used for computer programming. Language is universal to all humans and has a biological basis. Human beings have the capacity to produce an infinite set of utterances from a finite set of elements and they are unrestricted in what they can talk about.

Whether humans are born with an innate ability to process speech or whether this is something acquired through learning after birth is still a matter for debate. For some, language can only develop in a social setting. Humans acquire language through social interaction in early childhood. Nevertheless, the most current research suggests that babies develop language skills while still in the womb in response to their parents' voices. For B.F. Skinner (1904 – 1990) or Jean Piaget (1896 – 1980), language is a learned skill through a succession of trials, errors, and rewards for success. In other words, children learn their mother tongue by simple imitation, listening to and repeating what adults says. According to this view, when children come into the world, their minds are like a blank slate. Language acquisition results from simple interaction with the environment. For Naom Chomsky (1928-), linguistic faculties have less to do with natural selection. The environment only shapes the contours of this network into a particular language. For Chomsky, language learning is facilitated by a predisposition that our brains have for certain structures of language. In other words, the reason that children so easily master the complex operations of language is that they have innate knowledge of certain principles that guide them in developing the grammar of their mother tongue. In Chomsky's view, all languages share a "universal grammar" or a set of syntactic rules and principles. That is why even before the age of 5, children can, without having had any formal instruction, consistently produce and interpret sentences that they have never encountered before.

Why do humans undertake linguistic behavior? Language is not a mere string of words. Language embodies meaning well beyond the immediate and lexical expression. Aristotle declared that language is the representation of the experiences of the mind. For Jacques Lacan (1901 – 1981), language is central to human experience. He contends that the unconscious is structured like a language.

Language is thought to have originated when early hominids realized shared intentionality for communication and cooperation. Language facilitated hunting parties, warning of danger, or communicating with sexual partners. Babies cry when they are hungry, uncomfortable or unattended to. Language facilitates human beings to express themselves and manipulate objects in the environment. Language is learned. Through the use of language, any skills, techniques, products, modes of social control, and so on can be explained, and the end results of anyone's inventiveness can be made available to anyone else with the intellectual ability to grasp what is being said.

Languages evolve over time. Linguistic change is a universal process. Languages live, die, move from place to place, change and diversify with time. In the past, physical barriers such as oceans, high mountains, and wide rivers constituted impediments to human intercourse. However, modern technology in the fields of travel and communications has made such geographical factors of less and less account. Political restrictions on the movement of people and of ideas have also been rendered unnecessary. Language diversity distorts connectedness. The new trend is the evolution towards linguistic homogeneity. Slaves working on American plantations were forced by circumstances to acquire pidgin languages to share their experiences. Just like pidgin languages later evolved into creoles, speakers of diverse languages are slowly but surely evolving towards a common language to be understood by all living things and objects alike. In the space of one or two centuries if current trends continue uninhibited a perfect language (a universal language) is likely to emerge to facilitate international intercourse and cooperation, media, commerce, legislation, instruction etc.

The world is witnessing prolonged and regular contact between speakers of one or more languages, a process that regularizes the mixing of languages. The result is something akin a *lingua franca* (or working/bridge language). The perfect language may not necessarily be today's *lingua francas* such as English, Arabic, Chinese, Russian, French, Spanish, Swahili etc. but rather a mix from the more than 7000 languages on the planet. Linguists are torn between linguistic homogeneity and language diversity. However, linguistic homogeneity is decisive principle in the contemporary world. The driving principle is the quest for universal connectedness. Language is the means through which humans share their desire and attempts to overcome mortal life.

Underlying *The Perfect Theory* is the understanding that human beings learn languages, produce and understand utterances because they want to communicate their experiences of the surrounding world. Language is one among many human adaptive capacities. Language gives humans power in relation to their environment. When people speak mutually unintelligible languages they are unable to understand each other. People continually adopt and modify the language of choice to suit their particular communication needs and experience. Languages are going extinct to enhance contact and exchange. Human beings use language to share their experience as mortal lives with the hope of controlling their fate.

Human beings have needs and wants. Economic systems structure and guide production, distribution, and consumption of goods and services in an economy. The economic system also deals with the allocation of value. The economic system is composed of people, institutions, rules, and relationships. Imperfect economic systems often fail to attain optimality and are instead inefficient. They rationalize why not all of society's goals can be pursued at the same time by anchoring their argument on notions of scarcity - the idea that the society has insufficient productive resources to fulfill all human wants and needs necessitating trade-offs. Sub-optimal outcomes associated with imperfect economic systems include unemployment (forced idleness or brain waste), business cycles, the debt crisis, famine, poverty, malnutrition, brain drain, capital[6] flight, inflation, deflation, stagflation, recessions and violence/war. On the whole, imperfect economic systems cannot guarantee life in perpetuity.

---

[6] Capital is simply monetized existing wealth

Adam Smith thought that the shopkeepers at the corner shop flaunt merchandize at the bus stage because s/he is motivated by the pursuit for profit. Nonetheless, profit is not an end rather a means to an end. The marketplace is the point of connectedness. Money is merely the pricing mechanism. Money serves as the payment and settlement system to facilitate human interaction. The universe is structured such that one has to serve others (provide goods and services) to maintain life. Individual ownership (and scarcity) and multiple currencies are adaptive challenges that have a solution in public spaces (connectedness). In a situation of abundance (a perfect economy) there will be labour mobility/brain circulation thus full employment; a single world currency thus inflation is manageable (in fact non-existent); free trade; global connections, networks, partnerships and integration.

Besides infirmity, the most common causes of death are injuries from accidents, murder, suicide, and execution. Dying is not part of life. We don't have to die. The universe can sustain us and the trillions of people that came before us. We don't have to make way for those who are yet to be born. The universe is and has been expanding. Life can be endless.

Since the dawn of humanity, human beings have desired to live forever. This truth is evident in mystical and religious pursuits, art, literature, fables, allegory, myths, medieval alchemists and current-day science. Human beings attain a fleeting glimpse of immortality in art, literature, scenic beauty, music and in heightened emotion especially love. Scientific advances have allowed humans to profoundly alter their environment in order to sustain life. Improvements in food supply, hygiene, medical care, and innovations have resulted in far fewer deaths caused by trauma, infections and epidemics. Scientific progress is also responsible for the increase in longevity. Today, ageing can be slowed, stopped or reversed by anti-ageing products and services. Advances in medicine (diagnosis, treatment, and prevention of ailment) have allowed doctors to more effectively remedy many diseases and external injuries that might have caused people to die prematurely. Overall, though, the average life spans are increasing, we are not satisfied with extending our lifetimes since we never will to die. Scientists grow and expand human knowledge to sustain life. Science, in the holistic sense, is the adaptive capacity to sustain life everlastingly.

CHAPTER FOUR

CONCLUSION

The attempt to discover the theory of everything has been the reserve of natural sciences particularly theoretical physics. This is not the case anymore. *The Perfect Theory* challenges the notion that space and time are the only elemental components of reality. *The Perfect Theory* presents a believable account of the nature of the cosmos demonstrating why the universe is best understood in terms of spatial and non-spatial dimensions. The universe is made up of matter (space) and twelve (12) non-spatial properties of matter that include gravity, electromagnetism, strong nuclear interaction, weak nuclear interaction, dark matter, dark energy, mental faculties, senses, emotions, time, life and death. Matter is anything that has mass and takes up space. Space and non-spatial relations are intimately linked. Non-spatial properties come in multiple forms and can be converted from one form to another. Everything in the universe is connected. Nothing is solitary. According to connectedness, the universe is connected as one entity in physical and nonphysical realms. This formulation could help elucidate almost all known phenomena in a single theoretical framework. Connectedness represents the collapse of boundaries and divisions to realize global integration in the quest to protect life. Humans and other life-forms evolve, practice, study and manipulate phenomena (non-spatial dimensions) to stay alive.

BIBLIOGRAPHY

Adebayo Adedeji (1999). *Comprehending and Mastering African Conflicts: The Search for Sustainable Peace and Good Governance*. London and New York: Zed Books.

Allen, Thomas George (1974). "The Book of the Dead or Going Forth by Day: Ideas of the Ancient Egyptians Concerning the Hereafter as Expressed in Their Own Terms". *SAOC* vol. 37; University of Chicago Press.

Anderson Benedict (1991). *Imagined Communities: Reflections On the Origin and Spread Of Nationalism*. London: Verso.

Aristotle (2006) [350 BC]. *Politics*. Trans. William Ellis. Echo Library.

Aristotle (2007) [350 BC]. *Metaphysics*. NuVision Publications, LLC.

Augustine (1958) [AD 354– 430]. *City of God*. Image Books.

Bell Stewart John (1987). *Speakable and Unspeakable in Quantum Mechanics*. Cambridge: Cambridge University Press.

Bernal Martin (1987). *Black Athena: The Afroasiatic Roots of Classical Civilization* (1987)

Blockley, David (2010). *Bridges*. Oxford University Press.

Boas Franz (1922) [1911]. *The Mind of Primitive Man*. Forgotten Books.

Bohr Niels (1983) [1928]. "The quantum postulate and recent the recent development of atomic theory", *Nature*, 121,

580-89 (1928). Reprinted in *Quantum Theory and Measurement*. 87.

Calvin John (1995) [1536]. *Institutes of the Christian Religion*. Trans. Ford Lewis Battles. Wm. B. Eerdmans Publishing.

Cave Stephen (2012). *Immortality: The Quest to Live Forever and How It Drives Civilization*. Crown Publishing Group.

Clausewitz Carl von (2007) [1832]. *On War*. Trans. Michael Howard & Peter Paret. Oxford University Press.

Collingwood R.G. (1946). *The Idea of History*. Oxford: Oxford University Press.

Darwin, Charles (1859). *On the Origin of Species by Means of Natural Selection, or the Preservation of Favoured Races in the Struggle for Life*. London: John Murray.

Demarest Arthur (2004). *Ancient Maya: The Rise and Fall of a Rainforest Civilization*. Cambridge: Cambridge University Press.

Dickens Charles (2007) [1843]. *A Christmas Carol*. CSA Telltapes.

Einstein Albert (2010) [1915]. *Relativity: The Special and the General Theory*. Trans. Robert W. Lawson. Andras Nagy.

Euclid (2002) [330 B.C.]. *Euclid's Elements*. Trans. Sir Thomas Little Heath & Dana Densmore. Green Lion Press.

Evans-Pritchard (1951). *Kinship and Marriage Among the Nuer.* Clarendon Press.

Feuerbach Ludwig (1841). *The Essence of Christianity.* J. Chapman.

Foucault Michel (1978). *History of Sexuality.*

Fried Morton (1972). *The Notion of the Tribe.* Cummings Pub. Co.

Fritjof Capra (1975). *The Tao of Physics: An Exploration of the Parallels Between Modern Physics and Eastern Mysticism.* Boulder, Colorado: Shambhala Publications.

Fukuyama Yoshihiro Francis (1992). *The End of History and the Last Man.* New York: Free Press.

Fyodor Dostoyevsky *et al.* (2004) [1880]. *The Brothers Karamazov. Barnes & Noble*

Galileo Galilei (1967) [1632]. *Dialogue Concerning the Two Chief World Systems.* Trans. Stillman Drake. University of California Press.

Gellner Ernest (1983). *Nations and Nationalism.* Oxford: Blackwell Publishing Ltd.

Gibbon Edward (1776). *The History of the Decline and Fall of the Roman Empire.* London: Henry G. Bohn.

Gress David (1998). *From Plato to NATO: The Idea of the West and Its Opponents.* New York: Free Press.

Grotius Hugo (2004) [1625]. *On the Law of War and Peace.* Kessinger Publishing.

Grotius Hugo *et al.* (1609) [2001]. *The Freedom of the Seas, Or, the Right Which Belongs to the Dutch to Take Part in the East Indian Trade.* The Lawbook Exchange, Ltd.

Gurr Ted Robert (1994). *Minorities at Risk: A Global View of Ethnopolitical Conflicts.* United States Institute of Peace Press.

H.G. Wells and Cook Paul (2008) [1895]. *Time Machine.* Arc Manor LLC.

Hayek Friedrich (1944). *The Road to Serfdom.* Routledge.

Heather Peter (2006). *The Fall of the Roman Empire: a New History of Rome and the Barbarians.* Oxford University Press.

Hegel Georg Wilhelm Friedrich (1994) [1807]. *Phenomenology of Spirit.* Indiana University Press.

Heidegger Martin (1978). *Being and Time.* Wiley-Blackwell.

Hobbes Thomas (2010) [1651]. *Leviathan.* Broadview Press.

Hume David (2010) [1739]. *A Treatise of Human Nature.* Trans. Jonathan Bennett. Accessed at: http://www.earlymoderntexts.com/authors/hume.html.

Huntington Samuel (1996). *The Clash of Civilizations and the Remaking of the World Order.* Simón & Schuster.

Jervis Robert (1976). *Perception and Misperception in International Politics*. New Jersey: Princeton University Press.

Kant Immanuel (1954) [1788]. *Critique of Practical Reason*. Forgotten Books.

Kant Immanuel (2010) [1781]. *Critique of Pure Reason*. Trans. Jonathan Bennett. Accessed at: http://www.earlymoderntexts.com/authors/kant.html.

Kant Immanuel (2010) [1795]. *Perpetual Peace*. Trans. Jonathan Bennett. Accessed at: http://www.earlymoderntexts.com/authors/kant.html.

Keohane Robert and Nye Joseph (1977). *Power and Interdependence*. Longman.

Keynes John Maynard (1936). *The General Theory of Employment, Interest and Money*. MacMillan.

Khaldūn Ibn (1974) [1377]. *Muqaddimah: An Introduction to History*. Princeton University Press.

Khamala Geoffreyson (2009). "Gender Dimension of Ethnic Conflicts in Kenya: The Case of Bukusu and Sabaot Communities". MA Thesis, Kenyatta University, Kenya.

Kovacs Maureen Gallery (1989). *Epic of Gilgamesh*. Trans. Maureen Gallery Kovacs. Stanford University Press.

Kurzweil Ray and Grossman Terry (2005). *Fantastic Voyage: Live Long Enough to Live Forever*. Rodale.

Lanza Robert and Berman Robert (2009). *Biocentrism: How Life and Consciousness Are the Keys to Understanding the True Nature of the universe*. BenBella Books, Inc.

Locke John (1821) [1689]. *Two Treatises of Government*. Whitmore and Fenn, and C. Brown.

Lovelock, James (1995) [1988]. *Ages of Gaia*. Oxford University Press.

Machiavelli Niccolò (2010) [1532]. The Prince. Trans. Jonathan Bennett. Accessed at: http://www.earlymoderntexts.com/authors/machiavelli.html.

Mackay, Alan Lindsay (1991). "Archimedes ca 287–212 BC". *A Dictionary of scientific quotations*. London: Taylor and Francis.

Malthus Thomas Robert (1798). *Essay on the Principle of Population: A View of Its Past and Present Effects on Human Happiness*. Forgotten Books.

Margalit Avishai (2004). *Occidentalism: The West in the Eyes of Its Enemies*. New York: The Penguin Press.

Marx Karl (1867). *Das Kapital: A Critique of Political Economy.* Washington: Regnery Gateway.

Mbiti John (1969). *African Religion and Philosophy.* Heinemann.

McTaggart J. M. E. (1908). *The Unreality of Time.* Blackwell.

Mearsheimer John J. (2001). *The Tragedy of Great Power Politics.* W.W. Norton & Company.

Mill John Stuart (1848) (2006). *Principles of Political Economy.* Cosimo, Inc.

Mueller John (1997). *Quiet Cataclysm: Reflections on the Recent Transformation of World Politics.* Zip Publishing.

Newton Isaac (2005) [1687]. *Principia.* Running Press.

Nietzsche Friedrich Wilhelm (2001) [1882]. *The Gay Science.* Cambridge University Press.

Ogot BA (ed.) (1976). *Zamani: A Survey of East African History.* London: Longman.

Phillips, Alban William Housego and Robert Leeson (2000). *A.W.H. Phillips: Collected Works in Contemporary Perspective.* Cambridge University Press.

Plato (1928) [380 BC]. *Phaedo.* Trans. Benjamin Jowett. Forgotten Books.

Plato (1954) [380 BC]. *Parmenides.* Forgotten Books.

Plato (2009) [380 BC]. *Timaeus and Critias.* Trans. Benjamin Jowett. Digireads.com Publishing.

Plato (380 BC). *The Republic*. Trans. Benjamin Jowett. BompaCrazy.com.

Plato's (1991) [399 BC]. *Euthyphro*. Ed. C. J. Emlyn-Jones. Cathal Woods.

Popper Karl (1936). *The Poverty of Historicism*. Routledge & Kegan Paul.

Popper Karl (1945). *The Open Society and Its Enemies*. Routledge.

Ricardo David (1817). *Principles of Political Economy and Taxation*. John Murray.

Rousseau, Jean-Jacques (2010) [1762]. *The Social Contract*. Trans. Jonathan Bennett. Accessed at: http://www.earlymoderntexts.com/authors/rousseau.html.

Sagan Scott Douglas and Waltz Kenneth (2002). *The Spread of Nuclear Weapons: A Debate Renewed*. Norton.

Said Edward (1978). *Orientalism*. Vintage Books.

Said Edward *(1981)*. *Covering Islam : How the Media and the Experts Determine How We See the Rest of the World*. Random House.

Schneider S. Michael (1995). *A Beginner's Guide to Constructing the universe: The Mathematical Archetypes of Nature, Art and Science*. HarperPerennial.

Schopenhauer Arthur (1818) [1958]. *The World as Will and Representation*. Falcon's Wing Press.

Schrödinger Erwin (1944). *What Is Life? Mind and Matter*. U.P.

Service Rogers Elman (1975). *Origins of the State and Civilization: The Process of Cultural Evolution*. Norton.

Shakespeare William (2006) [1603]. *The Tragedy of Macbeth*. Echo Library.

Smith Adam (2011) [1776]. *An Inquiry into the Nature and Causes of the Wealth of Nations*. Cosimo, Inc.

Spengler Oswald (1918). *Decline of the West*. Oxford University Press.

Spivak Gayatri *Chakravorty (2010) [1988]. "Can the Subaltern Speak?"* In *Marxism and the Interpretation of Culture*. University of Illinois Press.

Sun Tzu (2nd century BC) [2005]. *The Art of War*. Trans. Lionel Giles. El Paso Norte Press.

Thucydides [1946]. *History of the Peloponnesian War*. Trans. Sir Richard Winn Livingstone & Richard Crawley. Forgotten Books.

Tolstoy Leo (2012) [1869]. *War and Peace*. Trans. Constance Garnett. Random House Publishing Group.

Toynbee Arnold J. (1934). *A Study of History*. Ed. D.C. Somervell. Oxford University Press.

Tutu Desmond (2014). "Desmond Tutu: a dignified death is our right – I am in favour of assisted dying". http://www.theguardian.com/commentisfree/2014/jul/12/desmond-tutu-in-favour-of-assisted-dying. Accessed Sunday, July 13, 2014.

Waltz Kenneth (1959). *Man, the State, and War.* Columbia University Press.

Ward-Perkins Bryan (2006). *The Fall of Rome and the End of Civilization.* Oxford: Oxford University Press.

Weber Max (1968) [1919]. *Politics as a Vocation.* Fortress Press.

Weber Max (1998) [1905]. *The Protestant Ethic and the Spirit of Capitalism.* Roxbury Pub.

Yogananda Paramahansa (2008) [1946]. *Autobiography of a Yogi.* Diamond Pocket Books (P) Ltd.